DR. GOLEM

HARRY COLLINS & TREVOR PINCH

DR.GOLEM

HOW TO THINK ABOUT MEDICINE

THE UNIVERSITY OF CHICAGO PRESS • CHICAGO AND LONDON

HARRY COLLINS is distinguished research professor of sociology at Cardiff University, where he also directs the Center for the Study of Knowledge, Expertise, and Science. TREVOR PINCH is professor and chair of the Department of Science and Technology Studies and professor of sociology at Cornell University.

The University of Chicago Press, Chicago 60637
The University of Chicago Press, Ltd., London
© 2005 by The University of Chicago
All rights reserved. Published 2005
Printed in the United States of America
14 13 12 11 10 09 08 07 06 05   1 2 3 4 5
ISBN: 0-226-11366-3 (cloth)

Library of Congress Cataloging-in-Publication Data

Collins, H. M. (Harry M.), 1943–
    Dr. Golem : how to think about medicine / Harry
Collins and Trevor Pinch.
        p. ; cm.
    Includes bibliographical references and index.
    ISBN 0-226-11366-3 (cloth : alk paper)
    1. Medicine, Popular.   2. Medicine—Philosophy.
    [DNLM: 1 Sociology, Medical.   2. Consumer Par-
ticipation.   3. Delivery of Health Care.   4. Physician-
Patient Relations.]   I. Pinch, T. J. (Trevor J.)   II. Title.
    RC81.C695 2005
    616′.001—dc22

                                        2005001621

Contents

Preface and
Acknowledgments

In the two earlier volumes of the *Golem* series we set out our claim that the products of science and technology are better thought of as "industrial diamonds" rather than brightly polished jewels—science, as we explained, is often untidy and messy. Science is a golem. As we said in the first volume in this series:

> A golem is a creature of Jewish mythology. It is a humanoid made by man from clay and water, with incantations and spells. It is powerful. It grows a little more powerful every day. It will follow orders, do your work, and protect you from the ever-threatening enemy. But it is clumsy and dangerous. Without control a golem may destroy its masters with its flailing vigor; it is a lumbering fool who knows neither his own strength nor the extent of his clumsiness and ignorance.
>
> A golem, in the way we intend it, is not an evil creature but it is a little daft. Golem science is not to be blamed for its mistakes; they are our mistakes. A golem cannot be blamed if it is doing its best. But we must not expect too much. A golem, powerful though it is, is the creature of our art and of our craft.

To clear up a frequent misreading, it is not the danger of the golem to which we refer but its well-meaning clumsiness.

This claim about the clumsiness of science comes as less of a novelty in the case of medicine than in the case of physics and the like because death and illness are always with us, so we know that medicine is fallible.[1] The really hard question that remains is, "Knowing medicine is fallible, what should we do?" In the cases of science and technology looked at in our earlier volumes, we suggested that the biggest change needed was in perception. If the public knew how science and technology really worked, they would be better at making choices that turned on these issues, and this would eventually affect their lives, mediated, for example, by the way they acted at the ballot box. The difference in the case of medicine is that, as individuals, we do not have the luxury of waiting for "eventually."

To put this another way, the authors of the two earlier books in this series had two kinds of relationships with their material. In the bulk of the case studies, we redescribed episodes of science and technology where others had done the primary research; in some of the cases, however, we did the primary research ourselves. In this book we find we have a still greater intimacy with the material. In several of the chapters we find ourselves mentioning our own involvement in medical dramas, major or minor. We find ourselves discussing our own decisions about what to do in a way we never had to in the earlier two volumes. Indeed, the disagreement between the authors in the matter of vaccination, indicated in the last substantive chapter of the book, twice came within a whisker of ending the entire project, until we found a way of expressing two different views about medical choice within the one analytical frame. *Dr. Golem* has, then, been a much harder book to write than *The Golem* or *The Golem at Large;* the two earlier books were less involving—less direct. Here we have to decide what to *do* as well as what to *think*. And we could have used many more examples from our own experiences: the doctor whose view of treatment was so completely "scientized" that he treated diagnosis as a "fault tree," and never examined the patient himself (one of the authors) but sent him for an x-ray and looked at nothing but the technical report. That bad doctor put one of us through an agonizing summer before a pharmacist explained how to handle the excruciating symptoms of the not very serious underlying damage. We could have described the complete failure of orthodox medicine to cure a chronic bad back, misdiagnosed and wrongly treated with drugs, the resolution of which was begun with a simple manipulation by a chiropractor. We could have described the unreasonable pressure put on one of us to have a serious operation when mild medication proved sufficient to alleviate the symptoms if not the cause.

Generalizing from this kind of personal experience is dangerous, however. Because orthodox medicine represents the "official" view, it is its occasional failures that make the headlines, not its steady successes. For each of these cases of failure, there are many more where the pharmacist's advice or the alternative medical treatment would fail, whereas medical *science* would get it right. For each case of overenthusiasm for a radical intervention, there are one or more cases where any kind of intervention was wisely advised against—including one which would have put money in the pockets of a private consultant. Indeed, both authors and their families have benefited many more times, and just as strikingly, from their regular doctors' calm prognoses, sound recommendations, and even from the occasional display of brilliant diagnostic virtuosity.[2] Knowing, as all of us must, that even if we are lucky enough to have had few such experiences in our lives, the time will come when the failing body will start to ask serious questions, we have tried to find a way through the tangle of an imperfect medicine that can often work dramatically well. Just pointing to the problems would not be good enough. We have felt obliged, if not to find a solution, at least to provide materials and arguments that help us think better about issues.[3]

Partly as a result of this, there is far more independent writing in this book than in earlier *Golems*, it being almost equally divided between our own analyses of the sources and original research on the one hand, and exposition of others' work with a minimum of intervention on the other. The full bibliographic references to the works discussed both in this preface and the other chapters, as well as additional reading, will be found in the bibliography at the end of the volume. The first drafts of the introduction, chapters 1, 2, and 8, and the conclusion were written by Collins; the first drafts of chapters 3, 4, 5, and 6 were written by Pinch; and chapter 7 is a reprint from an earlier *Golem* volume. That said, each author has contributed to the other's work, and we share responsibility for what is here. The sources we use in this book have been more disparate than those referred to in the previous *Golem* volumes and this has meant more use of footnotes. The reason is that the social analysis of medical literature has fewer self-contained case studies written with the intention of exploring knowledge, and less of a tradition of in-depth fieldwork. On the other hand, the literature is vast and heterogeneous. These, then, are two other ways in which this has been a harder book to write than the science and technology *Golems* and why both our preface and introduction are longer than might have been expected: we have had to explain why we have chosen to concentrate on the tiny sample of the writings about social aspects of medicine that we have selected and still say something useful, and we have had to develop more of an argument and a structure. Here are the main sources.

Chapter 1 turns on Collins's reading of the literature on the placebo effect plus his own earlier writings on "experimenter effects." Collins used a wide variety of sources, most of which are listed in the bibliography, but began by reading the articles in Anne Harrington's edited volume *The Placebo Effect: An Interdisciplinary Exploration*.[4]

Collins based chapter 2 on an original research project carried out by Joanne Hartland and himself in 1994–95, "Bogus Doctors: The Simulation of Skill"—ESRC (R000234576)—with some additional material gathered from a survey of American media sources conducted for this book by Matthew Wong. Many passages of text found in the chapter are taken directly from early draft work prepared by Hartland with Collins's assistance.

Chapter 3 is based on Pinch's redescription of Michael Bloor's article on the difficulties of diagnosing the need for tonsillectomies. The latter part of the chapter, which deals with the different types of expertise needed in interacting with doctors, is Pinch's own work.

Chapter 4 on alternative medicine is based upon Pinch's reading of Evelleen Richards's monograph *Vitamin C and Cancer: Medicine or Politics?* We add our own conclusion, which differs from that of Richards.

For chapter 5 Pinch mainly used material from Robert A. Aronowitz, *Making Sense of Illness: Science, Society, and Disease* (especially chapter 1, "From Myalgic Encephalitis to Yuppie Flu: A History of Chronic Fatigue Syndromes"). Additional material came from an article by Jerome Groopman, "Hurting All Over," in the *New Yorker* magazine, and from Lee Monaghan's case study of bodybuilders in his book, *Bodybuilding, Drugs, and Risk.*

Chapter 6, which deals with the case of cardiopulmonary resuscitation, is largely based upon Pinch's redescription of chapters 2 and 3 of Stefan Timmermans's book on the topic, *Sudden Death and the Myth of CPR,* but, again, we add our own conclusion, which is not wholeheartedly endorsed by Timmermans.

Chapter 7 is reprinted (with minor stylistic changes) from *The Golem at Large.* It is based almost entirely on Pinch's reading of parts of Steven Epstein's book *Impure Science,* though here it has been given a short new introduction by Collins to show where it fits with the themes of *Dr. Golem.*

Chapter 8 is based on Collins's immersion in the British MMR debate, his discussions with colleagues at Cardiff University—notably Tammy Speers and Lindsay Prior, who were themselves researching the issue—a series of research seminars with additional members of Cardiff's School of Social Science's research community, and his interaction with the Pinches, treated as typical respondents interviewed about their vaccination choices, especially in the case of pertussis.

We are extremely grateful for the generosity and efforts of the authors who have allowed their work to be "golemized." In every case but one they were unstinting with the time spent on correcting our misunderstandings. Because this book is about a topic which is more personal and political than previously addressed in the *Golem* series, we have on occasion found ourselves departing from those authors' own conclusions and the recommendations stemming from their work, for which we refer readers to the original works.

We thank Elizabeth Toon, Lindsay Prior, Alex Faulkner, Jens Lachmund, Nick Hopwood, Dr. Adam Law, Les Vertesi, Chloe Silverman, and a series of anonymous referees for many helpful discussions or comments and for making sure our own excursions into the area of medicine were guided by the relevant literatures. We thank Matthew Wong for help in preparing the final manuscript. Our editor, Catherine Rice, offered advice based on an astute reading and endless enthusiasm. The final responsibility for mistakes in exposition, infelicities of style, and errors of judgment or analysis remains our own.

Introduction *Medicine as Science and Medicine as Succor*

Everyone gets sick, everyone dies. If medical science were perfect, there would be less illness and more choice about death. Even a medical science that could not reverse the ageing process ought to be able to prevent premature death from disease and injury. But it is worse than this; on the broadest view, medical science doesn't make much difference at all. As studies of the health of populations show, medicine as we know it does little to increase the average expectation of life; diet, hygiene, and lifestyle have a much greater impact. Medicine, then, can offer little in the way of an extension of the human species' lease on life. If medicine is so fallible, what follows? What should we do?

This is both an abstract question and one that is direct and urgent. How much of our taxes should go to medical research? Should we continue to give to that cancer charity? Is it sensible to spend so much on organ transplants when the money, spent on hygiene in developing countries, could save so many more? These are the big questions. The "small" questions are: What shall I do about this sickness or injury, which is hurting or killing me now? Will today's vaccination put the health of my children at risk?

Which of a range of treatments should I choose when each is claimed to be the only one that can cure me? Are my symptoms "psychosomatic" or caused by a "real disease"? Of course, a "small" question is about as big as a question can get when you are the person needing to ask it.

So as not to confuse the big and the small questions we have to remember that medicine is not one thing but two: medicine is a science, like other sciences, but it is also a source of succor—a source of relief or assistance in times of distress. The two faces of medicine often conflict. One dimension of that conflict is urgency: medicine as a *science* has to try to get things right however long it takes, but medicine as *succor* has to provide an answer here and now. A related dimension is the "unit of suffering": though the science of medicine does little for the life expectancy of populations as a whole, it still makes perfect sense for each of us as individuals to reach out for the succor that medicine might provide in moments of distress. In these cases it is not the long-term science of medicine that we need but the short-term fix or, at least, hope. Perhaps in the long term we will know so much about the science of medicine that the answers to the big questions and answers to the small questions will converge—we will explain what this would mean in chapter 1 and return to it in the conclusion—but in the meantime the big and small questions will often be in tension, each making its own sense within its own context.

The tension is there because hope can damage medicine as a science. Hope can demand a shift of resources away from activity that might lead to long-term advances into short-term but uncertain or bizarre sources of pseudo relief. The philosopher Blaise Pascal explained that one should bet on the existence of a deity because it cost little, while the cost of being wrong was eternal damnation. We can replace salvation with health in our medical Pascal's wager: it makes sense for the individual to stand alongside Pascal in the wager, betting on a personal cure however poor the odds, because the alternative is death; it makes equal sense for medical science to bet against the individual for the sake of a longer-term chance of increasing the collective good. This tension forms the backbone of the book: it is about medicine as science versus medicine as succor, or to put it another way, it is about the interests of the individual versus the interests of the collectivity or in still other words, it is about the short term versus the long term. The perplexed individual, we believe, should find it easier to make medical judgments if they understand these tensions and how they work themselves out.[1]

The Main Theme: Individual and Collective

If there is a keynote chapter, it is the first, on the placebo effect. The last two chapters, on AIDS cures and vaccination, may be thought of as reflections of the first. These chapters, 1, 7, and 8, exhibit the main theme in stark form, metaphorically as well as literally enclosing chapters 2–6. The placebo effect—the Latin is *placere,* to please—is the name given to the alleviation afforded by the administration of drugs and treatments that have no direct effect on physiology; fake drugs and treatments often cure just as well (or badly) as the real thing for reasons that are not much understood beyond the phrase "mind-body interaction." The placebo effect shows that, at best, medical science has only partial control over its subject matter. The placebo effect, then, can make people better, but, at the same time, it is a massive embarrassment to the *science* of medicine. And that illustrates the main theme in a nutshell! The placebo effect is in the way of the progress of medical science even though it is a source of succor for the individual sufferer. With the placebo effect one has, if one wants one, a rational explanation for the persistence of medical quacks and alternative medical regimes that seem to lie outside today's conception of the chains of biological causality: they work because they enhance the conditions that allow the mind to influence the deep processes

3

of the body. This is not to say that none of these alternatives work as advertised—the uncertainty of medicine leaves room for a once scorned treatment to drift into the realms of medical respectability while another drifts out—but let us concentrate on those that work through the mind. The more they are good for the patient, the worse they are for medicine as a science. This is because, assuming that spending on medicine is a zero-sum game, the more demand there is for alternative cures, the less political and financial support will there be for medical science as we understand it.

Of course, one day we may come to fit the mind's interaction with the body into the well-understood causal chains which a science must aspire to develop. If that ever happens, then another part of the tension between succor and science will disappear. In the meantime, the placebo effect presents us with many dilemmas.

One consequence of the placebo effect that will be discussed in chapter 1 is the need for the randomized control trial (RCT). In an RCT one randomly selected group of sufferers is given the drug or treatment under test while another is given a placebo, all parties to the test being unaware of which is which and who is who. Randomized control trials illustrate our main theme with great clarity. Suppose you are a patient in a randomized control trial for a drug designed to treat a life-threatening illness. Would you prefer to be in the control group or the experimental group? If you were an entirely public-spirited person you would not care—your sole concern would be with medical science and the long-term collective good. You would be happy to take part in an experiment that would prove whether or not the new treatment was a cost-effective way to safeguard the lives of future generations (and cost-effectiveness is the constant companion of the collective good). But if you were a less than purely public-spirited individual, you would rather be in the experimental group, because the new drug *might* have a better chance of saving your life than the placebo—it offers at best alleviation and at worst some hope. This tension worked itself through in tests for the effects of the AZT drug on AIDS sufferers in San Francisco. The study is described in chapter 7 of an earlier volume in this series, *The Golem at Large*, and reprinted, with a new short introduction, as chapter 7 in this volume.[2] The AIDS sufferers coped with the conflict by sabotaging the science: they shared drugs and placebos among themselves so that everyone involved in the experiment had an equal chance of receiving at least half a dose of the potentially biochemically active substance; this meant that the doctors had no way to know whether the drug was working. In that case citizens chose to act as cure-seekers rather than seekers after scientific truth. (Later they modified their position.)

A closely related problem is thrown up by vaccination. In most cases vaccination is both an individual and a collective good; the vaccinated individual is protected from disease and, if vaccination is widely taken up, the disease dies out and the whole population is protected. In this way smallpox has been eradicated from the world. But if the vaccination itself is dangerous (and most vaccines carry at least a tiny risk), then it is in the individual's interest for everyone else to be vaccinated, leading to protection through eradication (so-called "herd immunity"), without themselves taking any risk associated with the jab.[3] Parents can face an agonizing choice when a vaccination is believed to be potentially harmful while the campaign as a whole has the potential to eliminate a disease.

In Britain, around the beginning of 2002, some parents began to believe that the combined measles, mumps, and rubella (MMR) vaccination could occasionally induce autism. A single doctor with a handful of supporters had published a paper discussing the possibility, although the medical community at large denied that there was any evidence for the existence of such a link. A growing body of epidemiological studies, which deal with the risks to pop-

5

ulations, was also negative: that is, there were no changes in autism rates in populations coincident with the introduction of MMR nor any greater incidence in countries that used MMR compared with those that did not. Nevertheless, parents' worries were amplified by widely publicized accounts of children who began to exhibit symptoms of autism following the MMR vaccination. The difficulty is that there are bound to be such cases on purely statistical grounds: chance will ensure that the first appearance of symptoms of autism (the cause of which is unknown) will sometimes show themselves shortly after an MMR jab. Overall the number of parental reports of autism symptoms following MMR injections did not exceed what would be expected by chance, but this does not make the coincidence less dramatic or heart-wrenching for afflicted parents.

In such a case the logic of Pascal's wager tells the parent "Do not take even the smallest risk with the health of your child; even though the odds of the maverick doctor being right are millions to one against, avoid the vaccination." But if all parents follow this logic and act in their own interest, then diseases such as measles will become endemic. Since the risk to the long-term health of a child from catching measles is undisputedly greater than the risk associated with the MMR shot, what seemed to be in their immediate best interest would turn out not to be.[4]

This is a classic case of what is known in political and economic theory as "the Prisoner's Dilemma."[5] The solution is for everyone to act in the public good, even though, because of the statistical nature of epidemiological research, there remains a slight possibility that a vanishingly small number of children can become autistic as a result of the vaccination. To repeat, there is absolutely no evidence that this is the case; it is just that it cannot be ruled out absolutely (as negative hypotheses can rarely be ruled out absolutely in any branch of science).[6] Given the overwhelming balance of evidence in the case of MMR, it is not difficult to see that the right choice for parents is to vaccinate; in other cases the choice might be more difficult.[7]

The Second Theme: Interacting with Medicine

The second theme of *Dr. Golem* is the different ways we can interact with medicine. In the earlier *Golem* books we argued that the key to understanding our relationship to science and technology is to see them as bodies of expertise rather than as a combination of logic and fact. We compared the expertise of scientists with that of legal advocates, travel agents, car mechanics, plumbers, and the like. Medicine too is a form of expertise, and a medical consultation is an encounter with an expert.

The craft aspect of medicine can be found even in surgery. Human bodies exhibit enormous variation. Insofar as the human body is like, say, a motor car, it is like a car before mass production was introduced. The representations of the human body found in models or medical textbooks are simplified, stylized, and idealized. Surgeons cutting open a body do not just find this vein or that organ laid out as in the illustrations; they have to explore and map the bodies of subjects like unknown territories. Even skilled practitioners can fail to find their way about.[8]

Another difference between cars and animate bodies is that the latter are largely self-repairing entities. If left alone, most of the time, a living body will repair itself, and this makes the science of medicine very much more difficult in two ways. First, it is difficult to measure the efficacy of treatment because one never knows whether it is the last medical intervention that gave rise to the cure or whether the body would have cured itself anyway. Second, what results in a cure is most often not the replacement or repair of a broken part, but an intervention within the self-healing process. Even drastic surgery depends on the body healing the wound by itself. Since the self-healing process depends on many factors beyond the understanding and control of medical science and tech-

7

nology, knowing the cause of a failed intervention is very difficult even if the major causal chains are well understood.

Physiological variation, then, is only the start. Humans vary in their history, circumstances, mental state, and behavior. The placebo effect tells us that the intentional, the psychological, and the social settings will all be causal factors in the process of healing. The state of the patient's body will depend on what has been eaten, drunk, smoked, worried about, loved, and put in orifices, over a lifetime. We might say that, as a human life unfolds, the interaction of the initial genetic legacy and the life that has been lived gives rise to an almost infinite potential for variation, and this will interact with the repair process alongside the causes of the initial sickness. To understand the human body as well as we understand the motor car we will need to resolve the problems of the social and psychological sciences as well as the physiological.

For these reasons the relationship between doctor and patient involves the patient much more than, say, the relationship between mechanic and car involves the car. So long as the patient is conscious, the relationship is more like a visit to the hairdresser than a visit to the garage. When we visit the hairdresser we begin with an evenhanded discussion of the appropriate "treatment," and we describe the end-state we would like to achieve. As the encounter with the hairdresser draws to its close we look at the outcome in the mirror and discuss whether things have turned out well. The hairdresser cannot do a satisfactory job without taking into account the internal state of the "hairdressee"—the customer's desires—as well as his or her external state: "sick" hair is "cured" only when the discontents which the customer brought to the salon are resolved; the hairdresser can be sure about the patient's conception of the disease only if all agree at the end of the encounter that a cure has been effected. Sometimes, of course, a hairdresser will insist that he or she knows best irrespective of the customer's wishes, and this produces a tension that can be comical. In hairdressing we know that in such cases the hairdresser has exceeded his or her brief.

As in hairdressing, the doctor often has to rely on the patient to describe the symptoms of the illness, because only the patient knows them. Sometimes this is difficult because the patient may not be good at describing symptoms or may have a fertile imagination. Furthermore, the set of circumstances that led up to the disease, "the history," is going to be significant and, again, only the patient knows it. Finally, it may be the patient alone—the sole witness to his or her internal states—who can say whether or not a cure has been effected. Much more often than in hairdressing, doctor and patient will disagree about the seriousness and the state of the sickness. The nature of the interaction—the location of the boundary between medical expertise and the patient's self-diagnostic expertise—is, as the sociologist might say, continually under "negotiation."

Where the boundary is located depends on many things. It depends, for example, on the strength and interests of the parties, and this depends, in part, on the illness or injury. Thus the anesthetized subject of surgery is in no position to contribute further to the discussion. If the consumer is the unconscious victim of an accident, violence, or physiological trauma, then even initial discussions are impossible.

Historically, the doctor has gained in strength as medicine has become seen as a "science." Before the nineteenth century a sick person might purchase the services of an apothecary, or a midwife, or a surgeon, or a physician, but the specialist's knowledge was still hairdresser-like. Customers might not have been able to cut their own hair, but they knew what they liked. In medicine, patients monitoring their internal states could reasonably claim to know whether a few more leeches were in order or if it was time for a bit of cupping. The state of the urine might give special clues to the specially trained eye, but anyone could look at it and agree or

9

disagree. To find ways to categorize disease that could not be con-tested by the patient, doctors had to gain access to private realms.[9] For example, by cutting open a dead patient a doctor could dis-cover a cause for disease beyond dispute by the corpse; the live per-son had known only that there was pain, or fever, or whatever, not that there was an unusual lump on the intestine. Doctors who did autopsies, in sacrificing dialogue with their patients, attained in-creased authority. The use of special instruments had the same effect. The stethoscope, introduced in 1819, needed skill to use and skill to interpret. The stethoscope created a discourse that could be shared only between trained stethoscope users. The au-topsy and the stethoscope began to exclude the patient from the discourse that was to become medical science.

The new sciences of life and the huge technological complexity of aspects of modern medicine have tipped the balance still further in the direction of medicine as a science and the doctor as an au-thority. The high point was probably the first decades after the Sec-ond World War, when science as a whole seemed to rule unques-tioned. Since the 1960s, however, critics of medicine and a slowly maturing understanding of the nature of science have, to some ex-tent, smoothed the edges of medical arrogance. The question of the proper relationship between doctor and patient, the qualified and the unqualified, is once more opening up. Patient-doctor relation-ships in diagnosis are discussed at greater length in chapter 3.

As we showed in *Golem at Large,* expertise is not always signi-fied by formal credentials. In that volume we encountered sheep farmers with an expert understanding of their animals and a deep knowledge of the local ecology of their land. The same point is made in the case of the AIDS cures activists (chapter 7) who acquired sufficient medical expertise to influence medical re-searchers, leading to a change in the way clinically controlled trials were conducted.

A sideways look at the extent to which unqualified people can develop a useful level of medical expertise is found in chapter 2, where we describe the astonishing success of bogus doctors. Much of medicine is a matter of giving comfort; still more of it consists of skills that can be learned on the job; and most of these skills have been thoroughly mastered by the teams of nurses that support the doctor, whether real or bogus. For these reasons, such bogus doctors as are found out are nearly always uncovered not when they carry out some medical procedure incompetently, but when they fail in another aspect of their lives not directly to do with medicine.

The bogus doctor case also illustrates the main theme, the tension between individual and collective: counter-commonsensically, individual patients with many kinds of conditions might actually be treated better by an experienced bogus doctor than by a novice straight out of medical school. Yet rarely would one knowingly *choose* a bogus doctor, and no policy of support for bogus doctors could, or should, be officially promulgated. At the aggregate level more formal training is better than less formal training, and an efficient system of licensing is a good thing even though, at the level of the individual, it may not always be so.

As we will argue, one could say that the tension illustrated by the bogus doctor case has been resolved by training and licensing paramedics and even first-aiders. Such training legitimates the craft face of medicine, its potential mastery by the relatively unqualified, and the ability to learn a craft by on-the-job training. Chapter 6 deals with what are, arguably, crucial medical interventions by only cursorily trained people—the use of cardiopulmonary resuscitation techniques (CPR). In recent years the techniques have become embedded within the emergency service infrastructure in general—ordinary people learning the techniques of mouth-to-mouth resuscitation and the like—and with public spaces being equipped with CPR technologies such as defibrillators.

Interestingly, a historical overview of the introduction of these techniques shows that, as in the case of a number of other medical interventions, there is little evidence that CPR actually makes much aggregate difference when it comes to saving lives. So this case again illustrates our main theme: individuals who suffer heart attacks or arrested breathing would still want someone to try to save their lives using these techniques even if, overall, the odds are not good.

We have explained that the very nature of human bodies and lives implies that there must nearly always be a high level of interaction between doctor and conscious patient. We have argued that the craft face of medicine offers a choice between the experience and qualifications in those who treat us. Nowadays, with our better understanding of the nature of expertise, there are also more systematic choices to be made.[10] We can divide these choices into three levels. At the basic level the citizen might want to "choose between experts." The citizen can choose between medical experts within orthodox medicine by asking for a second opinion, or might look for an alternative treatment—chiropractic rather than surgery for a bad back, or acupuncture instead of antidepressants. We illustrate the dimension of this kind of choice in chapter 4, which describes an alternative treatment for cancer. Nobel laureate Linus Pauling teamed up with a Scottish doctor, Ewan Cameron, to propose treatment of cancer with massive doses of vitamin C. The treatment was tested in two deeply contested trials conducted by a team at the prestigious Mayo Clinic. We examine the debate between the Mayo Clinic on the one hand and Pauling and Cameron on the other. These tests were subject to the familiar "experimenters' regress," described in earlier *Golem* volumes. In the end, the verdict of medical science on this treatment was negative. But individuals might still want to try the treatment when all other hope is gone—there is enough slack in the statistics and methodology of the tests to make sense of such a decision, although, we argue, not enough to justify more *public* spending on vitamin C research.

Another kind of interaction open to the citizen is made possible by the rising levels of education and easy access to information provided by the Internet. Citizens can try to develop their own expertise and, as in the early days of medicine, enter into a more evenhanded dialogue with the doctor. Sometimes the patient's expertise can be at quite a high level (what has been called "interactional expertise"), as in the case of pertussis vaccination that we will discuss in chapter 8.[11] A danger is that patients can gain a false impression of how much they know, because sources of information such as the Internet may be very persuasive even when real knowledge is sparse. Furthermore, if a few hours spent reading documents were enough to make a person into an expert, there would be no need for medical schools and on-the-job training; apprenticeship is essential in any profession that involves an element of craft, not least medicine.[12] Nevertheless, this does not mean that *every* attempt to gain expertise is based on misplaced confidence. We will refer to this kind of interaction with the medical world as "trying to become an expert." This kind of expertise is developed almost without any self-conscious effort when a patient with a chronic condition, such as diabetes, develops a high level of technical understanding of his or her own physiology.

A third way of interacting, which we will call "trying to become a scientist," and which we describe in chapter 5, takes place when citizens band together to establish the existence of new kinds of illness not recognized by the medical profession. For example, we have seen Gulf War veterans try to establish the existence of "Gulf War syndrome." Here the veterans of the 1991 Desert Storm campaign contacted each other and discovered what they believed to be a common set of symptoms caused either by their use of depleted uranium tank shells, or the enemy's use of chemical weapons, or the cocktail of vaccinations given to them to reduce their vulnerability to chemical and biological attack. Another example is "chronic fatigue syndrome" (CFS) or myalgic encephali-

13

tis. Is, say, CFS just the normal tiredness and depression from which we all suffer when things are not going well for us, or is it something that is caused by a virus or some such and should be counted as a definable illness? Perhaps "repetitive strain injury" (RSI) is another case that falls between exhaustion (of part of the body), and illness. The stakes are high in terms of the psychological self-definition of patients, the role of medical science, and the right to financial compensation. In these cases organized bodies of the self-diagnosed are trying to intervene in order to have themselves defined as suffering from an illness rather than a more diffuse lack of ability to cope with the world.

These cases extend beyond merely "becoming an expert." If patients can establish such new forms of illness and, perhaps, associated new forms of treatment, we would have to say that they have developed what is known as "contributory expertise" as opposed to interactional expertise. In chapter 5 we encounter groups who do indeed profess to operate as "lay scientists" defining new sicknesses. We also look at a closely related group—bodybuilders— who work up enough knowledge of pharmacology and bodily reactions to manage, maintain, and assess different steroid drug regimes.[13]

Uncertainty

The background to *Dr. Golem's* primary and secondary themes is the uncertainty of medical science. Nowadays, that medicine is uncertain is not news, and the earlier *Golem* volumes have already made the point for science in general. For that reason we have only one chapter the main purpose of which is to illustrate medical uncertainty; that is chapter 3 on the diagnosis of tonsillitis, the prevalence of tonsillectomies as an intervention, and the process of diagnosis in general. But every chapter reveals medicine's uncertainty in passing. The placebo effect is at the heart of it; the argument over vitamin C, the question over the efficacy of CPR, the debates over the existence of new forms of illness associated with exhaustion, the difficulty of identifying bogus doctors through their treatment regimes, and even the doubts about vaccination policy, all illustrate uncertainty. The randomized control trial compared with the treatment of broken bones shows that even medicine's gold standard is really a celebration of what medical science does not know about causal chains within the individual body.

Given this, and the fact that medical science has little effect on average expectation of life, it is all too easy to fall into an antimedicine/antiscience reaction. But, as we try to show, this is not the right way forward. Medicine still offers succor, and medical science still offers long-term hope. Medicine must be doing something right, and we can see this with a simple example. If we were not impressed with the power of antibiotics we would not be so worried about the spread of antibiotic-resistant germs! The trouble with antibiotics is not a matter of scientific uncertainty. Antibiotics are overprescribed because we are impressed by their effectiveness. It is also medical science that explains why overuse of antibiotics is dangerous—unfortunately, we don't act on our knowledge. Here the science is solid, but individuals are still insisting on antibiotics for the treatment of viral diseases, where they are ineffective, and for minor diseases, where they would be better off giving the body's own immune system a chance to build its strength. This is not to mention the feeding of antibiotics to farm animals that is driven by economic interest. Science is not to blame here, it is ignoring science that is the danger. What we have tried to do in *Dr. Golem* is steer a course between the exposure of medical doubts and difficulties and the sensible use of medical expertise.

Our Choices

In writing this book we too have had to make choices. If our sole concern were to save or prolong the maximum number of lives in the most cost-effective way, the whole book could be compressed into a single paragraph. We would simply say that all the money we currently spend on medical science should be spent on disease prevention. In the developed world we would spend money on improving people's understanding of diet, the need for exercise, the harmful effects of ingesting certain drugs (notably tobacco), careless driving, and impulsive sex. To be still more cost-effective we would forget all about the West and put our resources into improving hygiene and diet in the developing world.[14] That recognized, we have chosen to address people like ourselves—the rich inhabitants of the developed world. We talk of how our taxes should be spent, how much support we should give to medical research of different kinds, and how we should choose a treatment in the face of conflicting information. We are analysts of knowledge, and our concern is medical knowledge and its relationship to the individual. Since medical science happens largely in the developed world, it is the developed world that is our concern.

Another choice we have made is to try to explain certain principles without thinking much about the economic and political contexts of even the developed societies. For example, the drug companies, who do a great deal of the medical research that gets done, have little to gain from expensively demonstrating the physiological potential of substances on which they cannot make a monopolist's profit.[15] Thus if some common substance, already so well known that it cannot be patented, is said to offer a better cure for a disease than an expensive new drug developed in the secure and private environs of a company laboratory, it is unlikely to be tested. Again, the purveyors of *alternative* medicine have everything to gain from having their products brought within the ambit of state subsidized medicine, or at least state approved treatments. There are also the pressure groups which have financial incentives for defining new classes of illness. Medicine is practiced in a legal framework and this too will affect diagnosis and treatment. We may also be sure that at least some people obtain what economists call "utility" (of a hard-to-measure kind) through proselytizing for regimes of treatment that satisfy their view of what is "natural" or "holistic." In short, medicine is embedded in what we might call a "magico-industrial complex," and that goes for some of the episodes we recount in this book. The magico-industrial complex is not our topic, however. Our topic is the making of medical judgments in the face of the uncertainties and tensions found within even the most well-conducted and unbiased of sciences. We have shown in the earlier volumes of the *Golem* series that even the very best of the sciences and technologies have difficulty resolving disputes and, as sciences go, medicine is more controversy-laden than most. The kind of uncertainty that is intrinsic to medicine provides us with enough problems for this book, and these are what we have chosen for our central topic.

The Hole in the Heart of Medicine
The Placebo Effect

There is a hole in the heart of the science of medicine. It is the placebo effect. The placebo effect is the technical name for the mind's power to heal the body without obvious physical intervention. Sometimes the effect is triggered by administering a fake drug, often in the form of a pill made from a chemically inert substance. Such a pill is known as a placebo— from the Latin "to please."

We say that the placebo effect is a hole in the heart of scientific medicine because every time a new drug or other treatment is tested it has to be run against the placebo effect. That is, the placebo effect is taken to be so powerful that unless the effect of the drug is compared to the effect of a placebo it is almost impossible to tell whether improvements in health are due to the biological effects of the drug or the psychological effects of the encounter with one or more of the medical personnel, their paraphernalia, and the "medicines" or other "treatments" they supply. What this means is that every time a new drug or treatment is successfully tested, the members of the medical profession effectively proclaim two things simultaneously:

1 They proclaim, "We are such skilful medical scientists that we can invent new drugs and treatments," and they back up this claim by testing the new treatment and revealing its positive effects.

2 They proclaim, "We are such poor medical scientists that we cannot understand how the mind and the body interact," and they reveal how poor they are by factoring out the effect of the mind in the only way they know how: by comparing the effect of their new invention with the effect of a fake version of it.

Furthermore, in spite of all the medical science that goes into the preparation of new drugs and treatments, on an embarrassingly large number of occasions the fake is just as good or better than the real thing.

The Placebo Effect and Its Near Relations

Unfortunately, the placebo effect and its near relations are a lot more complicated than the above paragraphs suggest. To know where medical science stands today with the placebo affect we have to undertake an excursion through a fascinating but disorientating hall of mirrors. We are going to have to distinguish between the "real placebo effect" and the "fake placebo effect" and navigate a way through expectancy effects and reporting biases. Here we go.

Experimenter Reporting Effects

Consider those who carry out the drug trials. The experimenters have certain hopes and expectations about how the trials will work out. Wherever the results of experiments are marginal, experimenters' "reading" of their results tends to be influenced by what they want to see. In the 1960s psychologists showed that this had a dramatic influence, threatening the whole basis of experimental work in their subject. But that was just an extreme instance of the unconscious reporting bias that is present in all sciences, including the physical sciences. In the previous volumes of the

Golem series we gave examples of the way that the results of physics and other experiments are contested and interpreted in widely different ways by competing scientists. Though the reasons for different interpretations can sometimes be varied and subtle, the aspect that concerns us in the case of drug trials can be referred to as "experimenter reporting bias." Reporting bias differs from the placebo effect because it is, as it were, the effect of the mind on the mind (the mind of the experimenter), rather than the effect of the mind on the body (the body of the experimental subject). Reporting bias does not change the body; it merely changes the extent to which the experimenters think the body has changed.

To some extent reporting bias can be avoided if the person who does the analysis of the results of an experiment does not know what outcome to expect. The analyst, in other words, should be "blind" to the meaning of the experiment; this is usually arranged by sorting the subjects into randomly assigned groups for treatment and placebo and keeping the key to the membership of the groups hidden from the analyst.

Patient Reporting Effects: False and True Placebo Effects

Now imagine that a drug is being tested for its effect on depression. Depression is a subjective state, and the effect of the drug is likely to be measured by some kind of report made by the patients: the patients will say whether the drug made them feel better or not, probably recording their changed feelings on a form that encourages them to go into considerable detail. Here is another opportunity for a reporting effect to enter the equation. If some patients believe they have been treated with a powerful depression-relieving drug, whereas others think they have been given a neutral substance, then there is likely to be a biasing effect on the way the patients report their feelings. If they think the drug will make them better, they are likely to think they feel better—even if the drug has had no physiological effect at all. We can call this *"patient* reporting bias," as opposed to *experimenter* reporting bias. If there is no actual physiological effect on the patient we can call this the *false placebo effect*.

Of course, if the patient expects the drug to produce an improvement in health, it might actually produce such an improvement, because the state of the mind—for example a state of relaxed optimism—can affect the state of the body. This is the *true placebo effect*. The true placebo effect will often be actually or potentially measurable by physiological changes, such as an increased level of endorphins—euphoria-inducing chemicals—in the brain, an enhancement of the immune system, or improved healing of an injury. It is hard to say whether an increase in mobility in, say, a case of arthritis, which follows from a decrease in felt pain, should be described as physiological (due to more endorphins), or psychological: the boundary between the two is not clear. The overall point remains, however, that the reports of patients given a placebo who believe it to be a potent drug can be affected by a false placebo effect (which is a true reporting bias), or a true placebo effect.

When Is the Subjective Objective?

Of course, in the case of an illness such as depression, reporting bias and the placebo effect are not easy to separate. For example, if, as a result of reporting bias alone, a patient in a depression trial thinks he feels better does this not mean that they actually do feel better? Isn't thinking you feel better actually feeling better even if there is no physiological evidence that you do? This is one of the problems of measuring the efficacy of psychoanalysis and the like where there are no physiological correlates of the progress of the treatment.

One might think this problem can be avoided in those cases where the effect of a treatment is measured more directly than by a self-report. For example, patient's lung capacity might be tested before and after a test by asking them to blow into a device; or the length of time patients can walk on a treadmill might be used as a criterion of success in treatment for the lungs, or some such. Even in the case of these tasks, however, expectations about one's own performance can have an effect on actual performance in the absence of underlying physiological change. The amount of effort the patient puts into inflating the bag or walking the treadmill is, as it were, a self-report on their confidence in the effectiveness of the treatment even in the absence of a real placebo effect.[1]

Expectancy Effects

To confound things further, experimenters and human subjects cannot be thought of as independent groups. In the 1960s psychologists showed that school pupils' performance, as measured by outsiders, was affected by the teacher's expectations. If a teacher expected students to do well, they tended to do better than if the teacher expected poor results, even if reporting bias was eliminated by blinding. In this case it was the subjects of the experiments who were affected by the teacher's attitude—encouragement leads to higher expectations in the pupil and higher achievement. Let us call this the "expectancy effect."

The expectancy effect will apply to medical treatments too. If the person who administers the treatment is evidently optimistic about its potential, the optimism will transmit itself to the patient, reinforcing both patient reporting bias and real placebo effect.

There are, then, four effects that can lead to a positive outcome in a test of what is, according to medical science, a physiologically inactive substance or treatment. The four effects are

1   experimenter reporting bias;
2   the false placebo effect—in other words, patient reporting bias;
3   the true placebo effect in which the mind affects the physiology of the patient; and
4   the expectancy effect of the experimenter on the patient, which enhances 2 and 3.

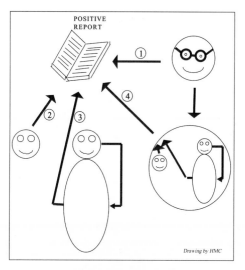

REPORTING BIAS (1)
FALSE PLACEBO EFFECT (2)
TRUE PLACEBO EFFECT (3)
EXPECTANCY EFFECT (4)

Fig. 1. Elements of the placebo effect. (1) Reporting bias, (2) False placebo effect, (3) True placebo effect, (4) Expectancy effect.

Because of these four influences, when human subjects are involved in experiments, both the subjects and the experimenters need to be "blinded." For example, in drugs trials, to avoid effect 2, the patients must not know whether they are taking the real drug or the fake; to avoid effect 4, those administering the experiment must not know whether they are providing the real drug or the fake to any one patient; and to avoid effect 1, those analyzing the experiment must not know which patients were administered real drugs and which fakes. When all these precautions are taken we have what is called a "double-blind" experiment—both experimenters and subjects are blind to the meaning of the experiment until it is over. Typically, in a double-blind test only after the effect of the treatment on each individual has been measured are the random codes which assigned subject to experimental and pla-cebo (control) groups revealed.

Physiological Efficacy

For the sake of completeness and clarity in the subsequent dis-cussion we must not forget that there is a fifth way in which a drug or treatment may affect the well-being of a patient: it may have the effect designed or discovered by medical science. We will describe this as a *"direct* chemical or physical effect," or sometimes, *"direct* physiological effect." This we will contrast with *"indirect* chemical, physical/physiological effects" which are the result of the mind's influence over the body even when this influence is mediated by something physical, such as an increase in endorphins in the brain or a boost to the immune system. In the classification above we can note that categories 1 and 2 involve no physical or chemical effects either direct or indirect, and category 3 involves indirect chemical or physical effects, which category 4 can boost.

Is the Placebo Effect a Fiction?

The placebo effect has been taken to be a scientifically estab-lished part of modern medicine at least since the 1950s. Studies suggest that between about 20 and 70 percent of patients seem to benefit from the administration of placebos. Perhaps most strik-ing is placebo surgery, where the appropriate anesthetic is admin-istered and an incision made in the skin, but there is no significant surgical intervention; this is reported to be highly effective. In-deed, the mock surgery sometimes seems to be more effective than the real surgery. For example, it appears to work for certain kinds of heart pain and back pain. In the mid-1990s, it was shown to work for arthritis of the knee; patients whose knees were merely stabbed recovered just as well as those whose knees were inter-nally scraped and washed out—the standard treatment that had been thought to be highly effective.

Unfortunately these seemingly straightforward findings are themselves contested. Now we have to traverse a still more disorientating wing of the hall of mirrors: unwell people can get better even if they have no treatment at all, and it is just possible that patients treated with placebos and those undergoing extensive medical intervention recover spontaneously at about the same rate. In other words, it could be that the patients given placebos or placebo treatments are not getting better because of the placebo effect, but recovering spontaneously, while the medical treatment is equally *ineffective* and the patients who undergo significant surgery are also recovering spontaneously. In this case, instead of there being a placebo effect which is as good as the real surgery, the placebo effect is no better than the real surgery and both are ineffective.

To find out that there really is a placebo effect, a different kind of experiment must be done. What must be compared are groups who are given a placebo on the one hand and no treatment at all on the other. For a placebo effect to reveal itself in these circumstances, the placebo patients would have to improve more than the untreated patients.

In 2001 two Danish doctors (Hrobjartsson and Gotzsche), analyzed all the articles they could find in which patients who were entirely untreated were compared with those given placebos. There are few experiments that are designed to test placebos directly, and most of the 114 trials the doctors examined had three groups: patients given medical treatment, patients given placebo treatment, and patients given no treatment all. They found that overall there was no significant difference between the placebo patients and the untreated groups in terms of improvement in the condition being treated.

That sounds decisive, and on first reading the report by the Danish doctors is convincing. The number of studies they analyzed and the number of patients involved were large. The study seems to overturn a huge raft of conventional wisdom. Careful examination of the cautionary remarks toward the end of the paper, however, makes the conclusion more resistible.

First, there are indications within the data of small effects of placebos on the experience of pain, and there is also room for large effects on some small subset of patients or conditions, if not all. These small effects and small numbers could easily be masked within the aggregate statistical approach taken by the Danish researchers. More worrying is the following rather complex piece of logic, the exposition of which will demand an increasing use of exclamation marks at the end of sentences.

A test of some treatment, placebo or otherwise, versus no treatment at all cannot be carried out blindly! Both the patients and those who are treating them will know who is not being treated; the fact that you are not being treated cannot be disguised or it would be not "no treatment" but, by definition, the administration of a placebo.

Now it gets complicated: if the doctors and patients know who is not being treated, one would think that would produce an expectancy effect and reporting effects that would make the difference between placebo and no-treatment patients even more marked than if the placebo worked! In other words, the nontreatment patients ought to feel pessimistic about their prospects, and those who are treating them ought to expect no improvement at all in that group, so one would think that there would be a strong reporting effect from both treaters and treated, and that this would be reinforced by an expectancy effect.[2] In sum, the crucial point is that even if there is *no* placebo effect, then in these nonblinded experiments there ought to *appear to be* a placebo effect because of negative reporting and expectancy effects on the untreated groups. In this *Alice in Wonderland* world this kind of experiment should not be able to fail! Whether it was really there or not, it should *appear* that there was a placebo effect!!

Now, since there was no apparent placebo effect in these experiments, there could not have been any expectancy and reporting effects, and this suggests there was something wrong with the experiments!!![3] Like Mendel's famous experiments on the inherited characteristics in peas, the results were so good that it looked as though they must have been false!

The Danish authors argue in reply that because in the majority of the experiments there were three groups, not two, neither patients nor analysts would have been concentrating on the difference between placebo and untreated patients and that this may have minimized reporting and expectancy effects. But this argument seems thin.

In any case, there is a quite different and opposite reason for distrusting the conclusions even if the absence of expectancy and reporting effects is not decisive. Inevitably, as we have argued, the untreated group would know they were receiving no treatment. Now, if their illnesses were serious, they might have decided that since they were getting no treatment within the study, they should treat themselves in ways which had nothing to do with the study (see chapter 4 for a similar claim in respect of vitamin C trials). This would not apply to the placebo groups, because they thought they were being treated. This difference in self-treatment could be responsible for the lack of difference in success rate between the groups.

Putting the two arguments against the Danish analysis together leaves us, as so often in difficult statistical sciences, not quite sure where we are, except that we know we cannot take the existence of a placebo effect as being quite as established as we once thought it was, but we are still far from being sure that it does not exist. What we need to settle the issue is a double-blind experiment between placebo groups and nontreatment groups—but we can't have one, by definition (and we'll have to end with yet another exclamation mark)!

Irrespective of the academic arguments, the drug companies, the agencies that administer the tests carried out by the drug companies, and the critics of the drug companies all treat the placebo effect as real. The critics point out that the so-called double-blinding does not work because patients can often guess whether they have been given the real thing or the placebo depending on whether the drug has any side effects, such as dizziness or a dry mouth. This means that even when a drug beats a placebo in a randomized control trial, it might be only because the drug, having side effects, also has a stronger placebo effect![4]

The drug companies and their agents treat the placebo effect as so real that they actually rate the placebo susceptibility of the patients they enroll in their trials. They try to avoid patients who are very susceptible to suggestion (covert psychotherapy) and the like.[5] And this is where we can leave the question of the existence of the placebo effect: in terms of its consequences for how we think about medicine, the placebo effect is real.

One More Complication

Consider tests of some drug or treatment that has long been known to be effective, such as hormone replacement therapy (HRT). Now suppose some doubts about its safety arise and it is thought prudent to test its efficacy again with a new double-blind control trial. In such a trial the patients, whether on the real drug or the placebo, have good grounds for belief that the drug has proven physiological efficaciousness. There is likely to be a very strong placebo effect because the patients will have a high expectation that if the drug they are taking is real, it will have a marked effect. In short, the strength of the placebo effect is partly a function of the strength of the patient's belief in the effectiveness of the real drug, and this, in turn, may be because of long experience by those who have taken the drug. In that case, if the trial reveals no difference between placebo and control group, this may be not because the real drug is not effective, but because its effectiveness has generated strong expectations in the placebo-takers. Any conclusion drawn from a negative result might, in these circumstances, be incorrect.[6]

Placebos and the Three Main Themes

The placebo effect certainly reveals the uncertainties at the heart of modern medicine. But it also poses a fascinating dilemma. If the placebo effect works, why not use it in a systematic way?

One answer to this is immediately apparent. Let us suppose you ask the patient, "Which would you prefer, real treatment or a placebo?" The patient has to say "real treatment," because as soon as you tell a patient it's a placebo, it's no longer a placebo; it is no treatment. Any attempt to offer a choice is self-defeating. (This is the logical complement of what we argued above: if you fool a patient into thinking that no treatment is a treatment, then it isn't no treatment; it's a placebo!) Doctors, however, may, and indeed do, dispense placebos in good faith so long as they keep it secret from their patients. A good doctor, wishing to offer some help in the case of an illness for which there is no scientifically accepted means of alleviation, should offer a placebo while concealing from the patient that there is no recognized treatment. But the efficacy of the treatment turns on the patient not being offered a genuine choice: "Would you like this placebo or not?" The doctor has to dissemble, and a dupe cannot make a genuine choice. The same applies to any agency that is responsible for the collective health of a population; placebos are a useful and important part of the armory of treatment, but "more placebos in medical treatment" is not something you can ask a population to vote for. Or can you?

Medicine as Science or Succor? Alternative Medicine and the Placebo Effect

Alternative medicine comprises all those cures, some traditional, some new, that are not recognized or supported, or rarely supported, by the main part of the medical science establishment. **29**

To define the borderline between orthodox and alternative medicine is difficult because the uncertainty in medical science leaves lots of room for the borderline to shift. For example, acupuncture is less readily dismissed than it was a couple of decades ago (See chapter 4 on alternative medicine).[7] Luckily, at this point we are not discussing the physiological potency of alternative medicine, and for the purpose of the analysis of the placebo effect we can simply assume that there is a subset of such treatments that have no physiological potency. Indeed, that is not much of an assumption—it is almost bound to be true. It is almost bound to be true because there are many orthodox treatments that have no potency, and it would be very odd if all the alternative treatments worked.

Let us call the subset of alternative medicines that have no physiological efficacy "empty treatments." We will make no attempt to identify them. The point is that even if empty treatments have no direct physiological power, a very large number of people think they benefit from them. Thus 42 percent of Americans use alternative medicine and so do 20 percent of British people. It is possible that all of these people gain nothing from the expenditure of such large aggregate sums of money except the certainty that they have left no stone unturned in their search for cures, but it is more likely that, at worst, they improve because of placebo effects, some of which will be mediated by real physiological change. Indeed, if the placebo effect is going to be significant anywhere, it is likely to be significant in the area of alternative medicine, which usually stresses care and optimism in respect of "the whole person" and almost never employs the cold and mechanical procedures that can be found in orthodox medicine. If the placebo effect cures, then alternative medicine, including the empty treatments, may be the best place for at least a proportion of the population to find it in its most concentrated and effective form.

And yet, the purveyors of alternative medicine resist the idea that its effectiveness is due to the placebo effect. They are like the medical profession proper in wanting professional recognition for the scientific, physiological, basis of what they do. And they *must* do this for the reasons we have described above; as soon as a treatment is announced as a placebo it ceases to be a placebo and becomes no treatment.

Let us go over this again by stepping outside the whole debate for a moment and imagining we could achieve an "Archimedean point" where we have the power to separate direct physiological efficacy from physiological change brought about by the placebo effect. Let us suppose that from this vantage point we can see that neither alternative medicine as practiced in Western societies nor the equivalents that are practiced in other societies, such as witch-doctoring, shamanism, voodoo, and the like, have direct physiological effects—they are all empty treatments—but that they can cure, often via indirect physiological changes, because of the placebo effect. In the case of "primitive" societies we can see that an outsider claiming that the rituals have no direct physiological effect would do little harm to their potency, since the medium of the cure is taken to be magical rather than chemical or physical. In large parts of Western society, however, the basis of sound intervention is taken to be chemical or physical. The placebo effect is fragile in our society just to the extent that our society is informed by a scientific worldview.

Here, then, is the tension between medicine as science and medicine as succor in a clear form. The state, in Western societies, generally leans toward medicine as science. For example, in today's British National Health Service (NHS) there is a growing emphasis on so-called evidence-based medicine. Drugs and treatments will be offered only if they have been proved in randomized control trials or something similar. But the very idea of a randomized control trial is an affirmation of the science-based way of life of our society, and, therefore, the discourse of evidence-based medicine is, by its very existence, something that reduces the effectiveness of placebo-based cures.

31

It is the view of the authors of this book that for a whole raft of reasons going well beyond medicine the scientific worldview is the one they want to see endorsed at the level of society, even though one has to accept that the health of at least some people will be indirectly damaged by it in the way just described. Here, then, is a way in which the tension between the individual good and the collective good is played out. An individual desperately seeking help for an ailment that orthodox medicine cannot cure and that alternative medicine cannot cure directly might be helped by taking advantage of an empty treatment via the placebo affect. To the extent that the government or any other agency responsible to the collective takes it to be its duty to foster a scientific worldview (and we think that is what such agencies should be doing), the chances of help coming from such a source are reduced. Governments can damage medicine as therapy even while improving it as science, but, nevertheless, they cannot choose to do otherwise.

The Scientific Gold Standard and Broken Bones

As explained, the existence of the placebo effect makes it necessary to use randomized control trials to test new drugs or other interventions, and the RCT has become the gold standard for scientific medicine. This, as we pointed out, has the ironic consequence that medicine's gold standard is itself a celebration of medical ignorance. We can show just what this means with a thought experiment.

Let us invent an ailment and call it Undifferentiated Broken Limb, or UBL. With UBL you know that one of the patient's four limbs is badly damaged, but you don't know which one. Let us imagine that someone invents a new experimental treatment for UBL and this treatment is a cast on the left leg—a CLL. We do a randomized control trial in which a control group is fitted with a cast around the neck to act as a placebo while each member of the

Drawing by HMC

**THE LOGIC OF THE SUCCESSFUL
RANDOMISED CONTROL TRIAL**

Fig. 2. The logic of the successful randomized control trial

experimental group is given a CLL. We can imagine that at the end
of the trial period when the casts are removed, up to one quarter of
the experimental group are much better, while there is very little
improvement among the control group. Thus the gold standard
test reveals that CLL is an effective treatment for UBL in about 25
percent of cases.

This victory for the randomized control trial shows us how little
we know about the body once we move away from some gross
assault such as a broken limb. Because we understand broken
limbs, we can see just how clumsy the randomized control trial is,
curing just a quarter of the population when better understanding
would lead to a more carefully tailored treatment for 100 percent
of victims.[8] To be in a position to understand all ailments in the
way we currently understand broken bones is what medical sci-

ence must aspire to. Such a complete understanding of the body (better, the mind and the body), would allow treatment to be tailored at the level of the individual cell (or individual thought, as it were), with as much certainty as it can now be tailored to the individual bone. When that happens the randomized control trial will disappear, just as it has in the case of bonesetting, and the main theme of this book will no longer be of interest, because medicine as science and medicine as succor—the long term and the short term, the interest of the collective and the interest of the individual—will have converged.

We do not know if such a state will ever come about—probably it won't, since, as we argue in the introduction, it would mean that the social and psychological sciences as well as the physiological had been perfected. But we cannot give up the hope that we will one day get there, and that is why we must hold on to medical science even though it is fallible in so many ways. In the meantime we can see that the very description of the RCT as medicine's gold standard means that the tensions that form our main theme remain and that each citizen will continue to have difficult choices to make. We hope the chapters, including this one, will reveal why maximizing short-term individual gain at the expense or in the face of science is not always the right, or even the best, choice.

Faking It for Real
Bogus Doctors

One way to understand the nature of a skill is to ask how hard it is to fake. We can learn a lot about faking and fraud from the newspapers, films, and television. The film *House of Games,* directed by David Mamet, took the viewer into a kaleidoscopic world where nothing turned out to be quite what it seemed. The techniques of confidence trickery portrayed in *The Sting,* with Paul Newman and Robert Shaw, were taken from journalist David Maurer's brilliant sociological analysis published in 1940, *The Big Con.* More familiar nowadays is the television show *Faking It,* where hamburger cooks are trained to take the place of gourmet chefs, punk rockers conduct a symphony orchestra, classical musicians take the role of the disk jockey in a club, and so on. The showdown comes when the faker competes with real chefs, classical conductors, and DJs in front of a panel of judges. Usually the expert judges are unable distinguish the fake performance from the real thing. Even assuming that what we see on the screen is not too distorted by the editing process, *Faking It* is still far from the world of confidence trickery as it is practiced, and we can learn less from it than from the older sources. For example, in one way the panel of judges have an advantage over the typical victim of a con—they know there is a something funny going on; in the case of real confidence trickery, and this includes the bogus doctors case, the victim does not know and often does not want to know that he or she might be the victim of a cheat. In this respect *House of Games* and *The Sting* come closer to the truth. Both films turn on one of the most important features of confidence trickery: the "mark"—the person who is to be conned— must really want the fake to be genuine; in *House of Games* the mark falls in love with the con artist; in *The Sting* the gangster who is to be fleeced is led to believe he is a partner in a scam to cheat the bookies. To understand the success of bogus doctors it is important to realize that everyone who knows the person will be made to feel a fool, and the routines of the medical setting will be thrown into chaos, should a trusted colleague, especially one of long standing, turn out to be a fraud. So while it would not be right to say that medical personnel want to be conned, the fact is that in most cases fakery will be the last thing they will think of should they encounter incompetence in their team; it is much easier and more natural to cover up, help the person out, and assume they will soon learn to handle the problem. *Faking It* fails to represent this important way in which those surrounding the faker help him or her to assume a new identity.

In other ways, however, the performer on *Faking It* has a huge advantage over the regular con artist. The faker is (a) trained by a team of specialists to carry off (b) a single performance in (c) well-understood set circumstances and is (d) surrounded by people who are willing parties to the act and will ignore inappropriate behavior or failures of technique that are not visible to the judges. Of these the training is the most significant. In one sense the performers are not fakes at all, since they have undergone a period of the best possible and most intensive training—the only unusual thing about it is that the training period is very short. So *Faking It* tests how much you can learn of a skill in a short time rather than how good someone is at faking. *Intensive Training It* would be a more accurate title for the program. Bogus doctors, in contrast, have to obtain their training surreptitiously.

Knowing that a single performance in circumscribed conditions is all that is required is another huge advantage; the faker does not have to know the skill well enough to carry off the fraud in the essentially unforeseeable circumstances of everyday professional life. This advantage is multiplied by the willingness of those surrounding the performer to ignore moment-to-moment failures in the ordinary flow of social interaction. As we will see, most bogus doctors are trapped precisely because of their inability to maintain the fraud under these more diffuse and extended circumstances.

Of course, an important variable in the ability to fake a skill is the degree of difficulty. Imagine that you—a musical naïf—has tricked your way into the seat of the solo violinist in an orchestra. Imagine the opening of a well-known piece developing and the conductor turning to indicate it is your turn to play. One note and you would be exposed! If, on the other hand, you were hidden among the rest of the violins, you might just get away with it by making some sawing motions without touching the strings and hoping everyone else was too busy to notice that one violin was missing. Even if they noticed you were not playing, you might feign illness and hope the rest of the team would cover for you. Imagine, on the other hand, you have never worked as a waiter in a diner but you lie to get the job; within an hour, or a day, as a result of careful observation and perhaps with some help from sympathetic colleagues, you would be up to speed, and your lack of experience would be beyond exposure. Some of the characteristics of virtuoso violin playing, ordinary violin playing, and waiting at table can, then, be understood by thinking about faking.

Finally, the difficulty of passing as a skilled performer depends on how well defined is the nature of an excellent performance. The untrained concert violinist is likely to be better at passing as a virtuoso when asked to perform a piece by John Cage than a piece by J. S. Bach. Indeed, we may turn back to film and see this feature of confidence trickery wonderfully exploited in Tony Hancock's 1961 portrayal of an avant-garde artist in *The Rebel* (retitled *Call Me a Genius* in the United States). In the film Hancock's incompetent daubs are, for a while, taken up by the Parisian arty set because it is so hard to tell the real thing from a fake when the very boundary of what counts as a skilled performance is the point at issue.[1]

How does medicine fare in this regard? The answer is that a lot of medical skills seem not too hard to fake, and we know this because there are a lot of bogus doctors about—around 10,000 in the United States, according to some estimates.[2] Of course, it is one thing to fake major surgery and another to hang up a shingle promising herbal remedies on the basis of no qualifications, and thus we don't know what the estimate of 10,000 really means. Nevertheless, we ought to be able to learn something about the nature of medicine by looking at the careers of certain kinds of bogus doctors. We can ask how long fakers managed to hold down medical jobs before they were discovered, whether and how they learned on the job, and how they were caught. In other words, is practicing medicine (of various kinds) like being a solo violinist, where the attempt to play the first note will expose you, or is it more like being a second violinist, or even a street musician or waiter? Our conclusions will be based on the analysis of multiple cases of bogus doctors in the United States and Britain. Let us describe a not untypical case to give a sense of the problem.

Unqualified anesthetist Abraham Asante worked in America. His unmasking came to be associated with his failure to notice that a patient had stopped breathing. But Asante was on his seventy-first operation at the time and had previously been highly praised by those who employed him. References written by army medical personnel for Asante included the following testaments: "At all times Dr. Asante has demonstrated the highest levels of medical knowledge, and the necessary skills to provide the medical service required of his medical section." "I would highly recommend Dr. Asante for positions of increasing responsibility." "I have found Dr. Asante to be a highly competent physician and a loyal member of this organization."[3] Asante, then, seems to have mastered all the skill necessary to carry out his chosen profession and had a long and successful career before he was found out. Asante, however, is untypical in one respect: usually, as we will see, the discovery of a bogus doctor has little to do with a medical mistake and much more to do with failure to behave in a way that is appropriate for a trained professional in some nonmedical area of life.

The North American Data

In a small survey we conducted we came across thirty-five reports of bogus doctors in newspapers in the United States, the first dated 1977 and the last 2004.[4] Sometimes these referred to bogus group practices rather than a single doctor. We found ninety-one cases reported in Britain in a more exhaustive survey of newspapers from 1966 to 1994. We carried out a much more detailed study of a few of the British cases.[5]

Medical fakery comes in different forms, many of which are of no interest for this chapter. For example, the U.S. reports include a homeless person who spent nights in a hospital pretending to be a medical assistant, four men who posed as doctors so that they could make intimate contact with women without any intention of treating them for any medical complaint, a man who tried to obtain medicines for his mother, another who pretended to be a doctor to get hold of narcotics, a woman who made money out of routine medical reports on potential bus drivers, a man who entered hospitals to steal wallets, one who posed as a doctor

39

to kidnap a baby, and various practitioners of alternative medicine who pretended to be qualified as orthodox doctors to boost their credibility. None of these are of interest here, since they did not try to make medical interventions for which they were not qualified—they did not try to "play the violin," as it were. Also included in the U.S. reports are doctors who were licensed to work in other places but not in the state in which they actually practiced. This certainly constitutes fraud, but it is not much of a *medical* fraud, since the perpetrators were medically trained to a fairly high level.

This last group draws attention to another problem: when bogus doctors are uncovered the tendency is to look for signs of medical incompetence and blame it on the lack of a proper license. In short, professional boundary maintenance tends to be confounded with medical incompetence—as though a violinist's poor playing were to be blamed on nonmembership in the musicians' union. If, then, a bogus doctor kills or injures someone, it is likely to be said to be a consequence of the bogusness, whereas if a qualified doctor kills or injures someone, as happens all the time, his or her training and qualifications will not be brought into question.[6] Thus, the U.S. reports include a fake plastic surgeon who butchered his patients, and the problem is attributed to his lack of training, whereas qualified plastic surgeons also butcher their patients from time to time without their qualifications being questioned. In the bogus plastic surgeon instance it may well have been the lack of training that led to the butchery, but one can see that there will always be a tendency to overemphasize any lack of qualifications where there is medical malpractice. This happened, for example, in the case of Abraham Asante described above. Though Asante performed well for many years and received glowing commendations from his colleagues, when he made his first mistake it was his lack of qualifications that was singled out for blame, whereas his track record suggests that this might have been the kind of accident that could happen to even the most well-qualified anesthetist.

Another complication is that in several of the cases the fakery was recurrent: a bogus doctor may be caught once but then return to the scam and work successfully for long periods. For example, in one well-known case, Gerald Barnes was initially unmasked following his misdiagnosis of diabetes, but subsequently he used his medical skills successfully in many places and for many years, including running a California medical practice where many of his patients were local FBI officers. The next time he was unmasked it was because he was recognized by someone associated with the initial case; this had nothing to do with medical incompetence. Between the first time he was caught and the second time he had treated many satisfied patients. If we were being very exact, we would count Barnes as two cases, one trapped as a result of medical incompetence and one caught out for nonmedical reasons. Here, however, we simplify matters by counting Barnes as a single case of bogusness associated with medical incompetence.[7]

Finally, bear in mind that the reports, in the nature of things, concern the less well-performing end of the spectrum of bogus doctors. Bogus doctors who perform well across the board are not caught and not reported. There may be many, many more bogus doctors who remain undetected.

With these reservations in mind we can examine the seventeen out of thirty-five U.S. cases where an attempt was made to fake medical skills. What is striking is that in only six out of the seventeen is it reported that medical harm was inflicted on any patient by a bogus doctor. Even for these six we cannot be sure that it was the harm that led to the actual unmasking of the bogus doctor; it may sometimes have been that some more prosaic reason for suspicion led to both the bogusness and the harm being brought to light at the same time. For example, in one of the British cases ("Atkins"—see below), it was a family member's anger that led him to go to police and reveal the long-running scam, and this in turn brought to light instances of inappropriate prescribing of medications. What we can establish from the breakdown of the U.S. reports is that it is not the case that fake doctors are quickly and easily uncovered or that they always do harm to their patients.

The British Data

The indications from the U.S. survey are more firmly established and more fully explained by the study of ninety-one British medical fakers who were reported between 1966 and 1994. Again, these were the bogus doctors who were caught; there may have been many more who never came to public attention, perhaps even the thousands who are said to exist in the United States. Of the ninety-one cases we know about, there were twenty-seven where the work of the fakers involved interaction with medical personnel. As in the American survey, the other cases involved using fake medical qualifications to aid some other illicit project such as trying to fool a bank manager, to con their way into someone's house, or to persuade someone to take off their clothes—frauds which shed no light on the nature of medical skills.

Once again, we might suppose that the majority of those who had worked directly with trained medical professionals would have been caught when they prescribed the wrong drug, messed up an operation, showed that they did not know how to diagnose a disease, failed to carry out a proficient medical examination, or the like. But this is not the case. We know how seventeen of the twenty-seven indicative British cases were caught. Three were caught because they could not manage ordinary life in a hospital sufficiently well to allow the rest of the staff to make excuses for them. The first of these asked a patient to pay privately for the removal of what the nursing staff could see was a harmless cyst. In the second the bogus doctor changed treatment instructions that had already been written up by other doctors. The third bogus doctor conducted an operation on a woman without her consent.

In another five cases the fakery was revealed as a by-product of investigations that had nothing to do with medical practice. One bogus doctor was arrested for bigamy, another for passport irregularities. Another was discovered as a result of making false insurance claims. The fourth exposed himself to suspicion when he took a part-time science teacher's job. The fifth revealed extraordinary scientific ignorance in respect of "cuckoo spit" during a casual conversation with a medical colleague!

Numbers nine and ten of the seventeen whose downfall we know about were recognized from other contexts. One tried to fake his way in a hospital where he had previously worked as a cast technician, and one was recognized as a schoolteacher when he came before a magistrate for a driving offence. Number eleven disappeared when he heard that a former colleague from abroad was about to pay him a visit and would be likely to recognize him. Number twelve out of the seventeen was betrayed by an angry member of his family. Number thirteen, an actor, admitted the subterfuge on a television chat show after he had left the bogus doctor profession. Number fourteen simply failed to return to his job after a holiday, leading to his history being uncovered. In the case of only the remaining three can we be certain that medical inadequacy was the prime cause of the investigation which eventually exposed them.

We can learn more about the flavor of day-to-day medical practice by describing five British bogus doctors for whom we were able to collect more details. One of these was interviewed directly, while a total of twenty-four acquaintances or colleagues of the five were also interviewed. As it happens, included among the five are two of the three who were caught as a result of medical mistakes. We will use pseudonyms for the five which start with the first letters of the alphabet—Dr. Atkins, Dr. Bailey, Dr.

Carter, Dr. Donald (who agreed to be interviewed himself), and Dr. Ferguson.[8]

DR. ATKINS

Four of the five bogus doctors acted out the role of hospital doctors. They each survived at least one medical job at a junior level before discovery. Let us start, however, with "Dr. Atkins," who worked as a general practitioner (GP). In Britain the GP is the first contact for patients. The GP administers general medicine, and refers patients on to specialists whenever things get complicated. This is a relatively easy role for a bogus doctor to carry off because he or she will spend relatively little time with other medical professionals and mostly will have to interact with the general public only.

Atkins arrived in Britain in 1961, and applied to the British General Medical Council (GMC) for registration, producing a copy of a medical degree and a reference from a hospital in Pakistan. His degree certificate was a copy of one awarded to a genuine doctor of the same name. Important distinguishing details on the certificate had been disguised, and his reference letter, it would turn out, was forged. Atkins had experience working with a "compounder" in India. A compounder is an unqualified chemist who mixes potions and remedies for common ailments. This may have given him a base of elementary knowledge from which to launch his ruse. Atkins was granted registration and set up his own surgery. He maintained his role as a doctor for about thirty years.

The thirty years did not pass uneventfully. A pharmacist who dealt with a large number of the prescriptions issued by Atkins had been surprised by the bizarre nature of some of the treatments called for. The most notorious prescription written by Atkins was for the antidandruff shampoo—Selsun—to be used to treat a throat infection. The pharmacist eventually decided to alert the local Family Practitioner Committee (FPC), an administrative body responsible for GPs, urging them to use their authority to investigate the source of these strange prescriptions. The FPC sent two qualified doctors to find out whether Atkins was physically and mentally fit to practice. One of these visitors explained how they had thought about the visit: "Two of us . . . sort of talked to him for half an hour and we couldn't see any sign of any mental illness and he wouldn't say he'd got any physical illness, so we had to report that we couldn't see any reason that he wasn't fit to practice." The other visitor said, "We came away with the belief that certainly there was no sign of any major mental illness at all. He gave, I suppose, tenuous explanations for prescriptions that had been passed, and that was it." The question of whether Atkins was a genuine doctor did not arise in these investigation. As was pointed out by one of the inspectors, "I think if . . . his credentials hadn't been examined in the first instance, there might have been some suggestion [about his authenticity], but his credentials were apparently good, and not only were they acceptable locally but they were acceptable to the GMC. . . . People ask the wrong questions. You say, 'How could this chap be so awful?' You don't say, 'Is this chap really who he pretends to be?' or 'Has he the right to be there doing what he is doing?'"

Thus, the bizarre regime of drug administration, including Selsun for a throat infection, was accepted by the two qualified doctors as within the envelope of the range of medical practices that can pass without being taken to demonstrate inadequacy. A retired consultant explained to us, "Medicine is not an exact science and people approach the same problem in different ways. If everybody was absolutely alike you could spot the one that wasn't alike. You see, everybody isn't alike, everybody is different. And provided the impostor is within the normal range, if you like—which is extremely wide—they don't stand out as an impostor."

One might think that Selsun was not exactly "in the normal range," but perhaps Atkins had stumbled on an unexpected property of the active ingredients in Selsun? Or, he might have been using it as a particularly dramatic placebo. Placebos are likely to work better if they have a nasty taste or other startling qualities. And the pharmacist who first called for an investigation explained, "I can look back on it now and say that it was obvious that this guy hadn't got a clue. But [at the time] I had no reason to think that he was a bogus doctor. He had been there for years and years. You would expect a bogus doctor to be caught out, or to be exposed, quite early on."[9]

Atkins's ultimate downfall came about only because a member of his family whom he had angered made a report about him to the Family Health Services Authority. The Atkins case nicely indicates the width of the envelope of what can count as reasonable therapy and the degree of satisfaction that an unqualified GP can give to his patients.

DR. BAILEY

Dr. Bailey had received some medical instruction in Afghanistan, but had failed medical exams there. He worked in a variety of hospitals in London. He took several temporary fill-in positions, known in the UK as "locum" posts, followed by posts as a casualty officer (known in the United States as working in the emergency room). He was granted temporary registration with the GMC in 1967 after producing a forged medical degree certificate from Kabul University. His fraud lasted for about three years.

During the course of his bogus career, several of his colleagues realized that Bailey had a lot to learn in the clinical arena, as his performance lacked assurance. But after initially contacting the GMC and being assured that his registration was in order, they took their job to be to assist and educate him, in the meantime diverting him from difficult cases. One member of staff suggested that Bailey "was almost the invisible man, drawing a salary for not doing very much."

Bailey was then supported by the team and might have gained substantial experience had he not been exposed by an anonymous caller who informed the GMC of his bogus application. He was arrested as he attempted to reenter Britain after a holiday when irregularities in his passport were noted.

DR. CARTER

Dr. Carter had experience as a student nurse and as a cast technician in various NHS hospitals. He obtained false certificates from an Australian medical school and was registered by the GMC in 1970. Carter worked in departments of surgery and anesthetics for about three years, progressing up the medical ladder and being offered a post at senior registrar level before he was uncovered. A consultant anesthetist who had employed this man as a junior doctor explained, "I've never been so shattered in my life as when a nurse came up to me and said the CID [the detective branch of the police] had been there . . . and I said, 'What for?' and they said, [Carter] 'had never qualified.' I felt as if I had been hit with an atom bomb." Or, as one of the nurses put it, "Had we been asked, and this was the general opinion of everybody when this came out, had we been asked to pick a doctor who was bogus, he would have been the very last of them all." A retired consultant who remembered working with Carter said his downfall "certainly wasn't [due to] a defect in his anesthetic practice."

Carter was arrested when an investigation of insurance fraud brought him to the attention of the police.

DR. DONALD

Dr. Donald produced a false GMC certificate when he took up his post. He was a failed medical student with experience working as a researcher in two different academic medical departments. He then worked for a year as a medical house officer in a district general hospital. Of this appointment, a colleague remarked, "I

47

had no inkling during the period he was working in the department that there was anything suspicious about his qualifications. . . . He must have done [his job as well as would be expected of a doctor at that level] because if anybody had raised the question of him not being able to cope then clearly we would have looked into that. . . . I know of nobody who complained about his manner or his behavior—his manner either clinically or as a person."

At the end of his year in general medicine, a Senior House Officer (SHO) vacancy arose in a prestigious dermatology department; their chosen candidate had been unable to take up the offer and this left the department short-staffed and looking for a locum. Donald covered the gap.

An unusual feature of life in his new department was that the SHO was required to accompany one of the consultants on two-person ward rounds without the support of other junior staff. The consultant explained the reason for this:

> I'm a bit old fashioned about how I relate to my junior staff and so on. I'm very loyal to them, and I'm quite generous to them and helpful and so on, but I think they have to earn your respect and so on. And I quite often put people a bit through the hoops when they come along so I can see what they're made of, and what level I can pitch my teaching at, and where I can help them and where I can't help them. And so, always on the first ward round or two, I'll put in quite a few little probes and see where they are . . . And I like to do one [ward] round a week really just on a one-to-one basis without anyone else.
>
> So I appeared on the ward on my first day back really from holiday, first time to meet him anyway, and he introduced himself.

According to the consultant, by the time the first ward round was concluded,

> It was obvious to me that this fellow really didn't match up to what he was supposed to have done. He'd been completely protected from the slings and arrows of the consultants on the big ward rounds but on a one-to-one basis he was naked, and there was nothing he could

do about it. I think that if you do a ward round with lots of people you will not pick up this sort of thing. It's a problem.

The other consultant on the dermatology team recounted a similar tale when he described his experience in his outpatient clinic: "I invited him in the routine course of events in the first week . . . to sit in on my consultation clinics, where it was one-to-one, consultant-to-patient, and no students. And obviously he sat alongside me, and when invited to examine or to comment on a particular case, it was quite obvious he knew very little indeed. I mean not of dermatology, which I wouldn't have expected, but of general medicine."

Within a few weeks suspicions were high enough to warrant an enquiry to the GMC. Donald was found to be using forged GMC registration documents, and was arrested.

DR. FERGUSON

Dr. Ferguson was granted limited registration with the GMC after producing medical degrees and references from institutions in the United States; he had previously worked as a paramedic.

Ferguson had completed the first section of an SHO rotation in a department of geriatric medicine. In this department he had managed to remain undetected. Here he was never faced with a completely unassessed patient—all the admissions to the department were referred either by another hospital or by their own GP, and they came with notes and ready-made assessments.

The second phase of the rotation was a particularly busy accident and emergency department, where the need to make on-the-spot individual diagnoses could not be avoided. The suspicions of his consultant were soon aroused because, as he explained,

> They [doctors in Accident and Emergency] are the first line. They will be taking patients off the street—it could be anything. It could be from a trivial little cut . . . or a patient in cardiac arrest; it could be a most traumatized patient. Of course, once they get hold of the pa-

tients then they can call for help, assistance of more senior people, you know, if it's a complicated case. . . . It is very acute—you have to make a decision. And he was the most stressed doctor that I have ever seen in my life. Sort of almost in a cold sweat. I mean doctors can be busy and they can be moody and they can be in a bad temper because they are just worked to the ground. There isn't any panic, just attrition, and worn out and disillusioned and all those sort of things which could push them to be bad doctors, but they will continue to be doctors. But he was just in a state of panic.

The consultant also recognized other problems:

He spelt wrongly even for an American, you know—I mean one accepts that "colour" is spelt "color," but when you meet people who can't even spell that, you begin to think that they probably haven't even had a basic education. . . . If this guy was talking to me I wouldn't know what he meant, he would talk about "a broken forearm"—it's very unusual for doctors to talk about a "broken forearm"; they would talk about either a Colles fracture or a Bennett's fracture or a Smith's fracture or a fracture of the radius and ulna. It's very unusual for them not only to talk about it, but to actually write down, "a broken forearm." There was something that made you feel uneasy. So that began to arouse suspicion. And whilst one was prepared to say "well after you get settled," and "let him sort himself out"—. [Then], he departed for the weekend and I just had a whole pile of everything that he saw, out on the table the next day.

That is, the consultant felt uneasy enough to undertake a complete audit of all the patients that Ferguson had seen. "I have to say that my suspicion at first was that he was probably a bad student and that's why he left the States; he couldn't cope with the system in the States, of course, because with writing like that he'd have ten patients suing him. And so then I started to think 'Perhaps he's got some cases against him in the States and he's decided to run off before he comes to trial.'"

The consultant's suspicions were aroused to the point where he undertook to investigate Ferguson's medical school and the origin of his references. The former was nonexistent, and the latter had been handed in by Ferguson himself. Yet this consultant admitted that Ferguson had not made any major medical mistakes—his problems were more to do with his manner—a state of panic—and his lack of basic educational skills. The consultant explained to us, "There weren't mistakes, in the sense that you treat a Colles fracture as if you were treating a Smith's fracture. But there were omissions—virtually every patient had large omissions. But not mistakes as such, in the sense of you were treating heart attack as indigestion. That is a mistake. But he didn't actually—I don't think he had lots of mistakes, he just had lots of omissions."

Thus, though we have classified this as a case where medical inadequacy led to exposure, the mistakes made by Ferguson were in medical etiquette or style or completeness of reporting rather than technique.

The case is complicated, because at his trial the charges against Ferguson included manslaughter. A patient under Ferguson's care had died after Ferguson had prescribed insulin for a chest infection. Was this a case of unambiguous medical incompetence? Four medical experts were called to give their opinions on the cause of death, but each offered a different opinion. One was sure that the insulin injection ordered by Ferguson had killed the patient, one of Ferguson's superiors disagreed, a third expert said that the treatment was inappropriate but irrelevant to the cause of death, and the fourth expert believed that the patient had died from an unrelated conditions—septicemia. The judge remarked, "There remains a vast area of doubt in this case and it remains wrong in my judgment for the case to go ahead." The jury was instructed to return a not-guilty verdict.

Learning on the Job

There can be little more frightening than to discover that a person in whom you have placed your trust, or even your life, is a faker—what a terrible revelation! Appropriately, the mass media report cases of bogus doctors in terms of shock and horror, as though every bogus doctor were an actual or potential killer. But a more nuanced look at the matter offers a different perspective. As we have indicated, the most remarkable thing is the small number of bogus doctors who are unmasked as a result of their medical mistakes and the extent of medical uncertainty that is revealed when charges of medical incompetence are investigated.

It turns out that there is enough variation and uncertainty in medicine to allow a fairly ignorant faker to enter the profession, to trade on the lack of knowledge of the public and the differences of opinion that are found even in the heart of the medical profession, and to survive for a while at least. As a retired doctor commented when discussing the case of Bailey, "On the whole we don't know very much about what makes people ill and well, and most people get well of their own accord. And this is not something that we boast about." Another retired consultant held the view that "the great majority of diseases that are cured, cure themselves, and most of the others kill you—they kill you in spite of the doctors— they may prolong life by a week or two or even a year or two, but eventually it will get the better of you."

The "looseness of fit" between sickness, diagnosis, treatment, and cure is recognized in the discourse of doctors. Writing in 1976, American sociologist Marcia Millman suggested that doctors justify one another's mistakes by claiming that each case was unique—a special example which the rules did not cover. The sociologist Charles Bosk, writing in 1979 about his research in American hospitals, drew the interesting conclusion that because medical practice was so filled with uncertainties, lapses in doctors' moral codes were more heavily punished than lapses in medical proficiency (though the situation may have changed with the increasing litigiousness surrounding the U.S. medical profession). More recent writings pick up the same themes. Marilyn Rosenthal, writing in 1995, reports interviews she conducted with

British and Swedish doctors exploring the theme of indecisiveness and uncertainty. She found that doctors do not like to use the term "mistake" or "error" in connection with their work, but prefer to think about "avoidable" and "unavoidable" accidents.

Crucially, even trained doctors first enter the medical profession from widely differing backgrounds and with very little on-the-job experience. As a result, the supporting team of nursing professionals and other doctors are unsurprised by initial displays of incompetence and are ready to lend a hand. The team will offer support, overlook minor mistakes, and rectify large ones, treating a new colleague as an apprentice rather than a fully equipped doctor. Thus, of nurses a consultant said, "The junior staff [doctors] don't do very much. . . . The casualty sister did most of it . . . if she thought that the junior was incompetent then she would help him do a dressing or lance a boil or things like that."

It is very easy for the supporting team to assume that an untrained doctor is simply a novice or someone from an unfamiliar medical regime—the rest of the "violins" fill in for the one poor player, as it were. Given the time and space afforded by the ready acceptance of mistakes and a wide range of practice within the medical profession, the competent bogus practitioner can learn on the job. This way the bogus doctor may develop enough experience-based competence to fool other professionals. Our interviews with those who had watched bogus doctors at work illustrate these points. A consultant orthopedic surgeon said, explaining why incompetence did not lead him to question a doctor's qualifications, "I had got eighteen years into my job and I had got used to having house surgeons who came knowing absolutely nothing about orthopedics at all. And indeed many of them knew very little about anything."

Again, one of Bailey's former colleagues told us,

He [Bailey] was very good about coming up to other doctors in the department and saying "could you help me I've got a problem here?"

And he did that frequently. And he would call you to see a patient and say "could you come and have a look at this girl, I'm a little unsure about what is going on and I'd like to talk to you about it." And you'd go in and say, well I think she might have appendicitis and he would say "oh yes that's what I thought but I wanted you to [look]." And so he did that quite a lot.

Of Donald it was said,

He would, in time, have seen other people, more experienced than himself, use the more sophisticated methods. And he might even have gone and said, "could I have a go at that?" or, "would you show me how you did that?" And most people would be flattered and say, "come along, I'll show you how it's done." So providing he started with the simpler things, and opted out if it was something too much for him, there is no reason why he shouldn't have improved. After all, we all learn our jobs largely by imitation. I mean, what you learn in medical school is of comparatively little use to you when you start on the wards. And when you start in anesthetics first of all, you would learn a great many things you can't really do because the only way you can do them is to start doing them with somebody holding your hand, and then go on until you are confident to do them yourself.

Talking about the same case, a different consultant explained the learning process as it had applied when she had taken charge of the junior staff in her department: "It always was a sort of graduated learning process whereby you did something and showed them, explained it, told them the whys and wherefores, and when not to, and how not to, and then you let them do it while you are there standing over them. And then my practice was that I would stand outside the door and watch, and at the least sign of distress I would go in and take over. He [Donald] was a quick learner, but he was well taught."

And "Dr. Donald" himself reported:

The actual house surgeon and senior house surgeon, when they come to the job first, they do very little. I was surprised at that later on when I was an anesthetist and I was watching. You could probably take a high school kid of eighteen and put him in this and give him a week's tuition and he'd probably function just as well as the house surgeon—who'd been trained for five years—in the role of assisting. But of course it is only a step, if you like, onto gaining more knowledge.

These passages bear on one of the themes of this book—the extent to which patients can turn themselves into experts, putting themselves into a position to discuss their ailments with medical professionals in an evenhanded way rather than simply accepting authoritative advice at face value. The point to note is that even after a formal medical training lasting many years, a junior doctor is still a novice. Patients, therefore, should not mistake even extensive book-learning for real expertise.

Given the fact that the team expects to support a novice doctor and that they will encourage learning on the job, it is less surprising that few bogus doctors are caught as a result of medical error. The two British cases who were uncovered as a result of their poor medical expertise were exposed in the very demanding situation of one-to-one interactions with experienced doctors. That said, even in the case of Ferguson the mistakes were subtle. It was his inability to learn how to handle himself in medical interactions that was the crucial problem rather than the actual treatments he provided—he was always in a terrible panic and he had not acquired the language of medicine. He would, as the consultant who trapped him said, "actually write down, 'a broken forearm,'" instead of using the appropriate technical term.

Though bogus doctors can survive in many different medical specialties (for example, bogus surgeons are not unknown), the risks vary. General practitioners often work alone, interacting little with other medical practitioners. Many of the patients of a GP will have standard chronic conditions which demand the doctor's sympathy, understanding, and ability to elicit a self-diagnosis (see chapter 3 on tonsillitis), more than her medical knowledge. This could account for the extraordinary patient endorsements which bogus doctors sometimes receive.[10] Even the hospital setting allows for some anonymity during those crucial first few months because of the regular turnover of staff. This gives a space for the supporting team to fill in the gaps and time for the faker to learn. Even when the bogus doctor is exposed in a one-on-one relationship with another specialist, the game is not necessarily at an end. In general, whatever the specialty and the circumstances, the more time the bogus doctor spends in medical surroundings, the less obvious will the difference between faker and qualified doctor become, both in terms of surface behavior and ability to carry out the job. Indeed, an experienced bogus doctor is almost certain to be a better doctor than a novice straight from medical school.

In every profession there is a spectrum of ability from the brilliant to the incompetent. It is tempting to think that all bogus doctors must be less competent than qualified doctors, but our analysis suggests that there will be considerable overlap. At the top end of the profession, the brilliant doctors will stand out clearly; at the bottom end of the profession, however, are the trained but incompetent, and the novices with little experience.

Figure 3 repeats, in diagrammatic form, what has just been suggested. The heavy line represents the competence of qualified doctors, while the light line might represent the bogus doctors— with the absolute number much exaggerated. What we are suggesting is that the mid-range of fakers, who have learned on the job, are more competent that the lower range of qualified practitioners, and that in that mid-range there will be a number who are practicing today in an unexceptionable way and have never been uncovered.

Bogus Doctors, the Individual, and the Collective

It now seems less strange to imagine occasions when a patient would choose a bogus doctor to continue the treatment even after

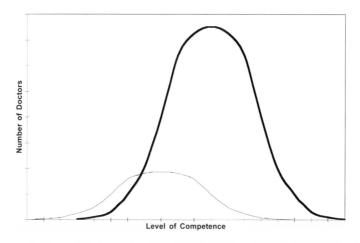

Fig. 3. Hypothesized competence of bogus doctors compared to trained doctors

he or she had been unmasked. For example, a patient might prefer to continue to accept treatment from someone like a Dr. Atkins than take the risk of placing her well-being in the hands of a qualified stranger. Atkins was a hard-working general practitioner of long standing with a reassuring bedside manner and a well-established member of the local community. If he had treated your family successfully for decades, why change?

This presents us with a puzzle. On the whole, experienced bogus doctors who are serious about learning on the job seem reasonably competent, so why are we shocked when they are uncovered? After all, in medicine everyone makes a few mistakes. The cynical answer is to say that it is all about the self-interest of the medical profession, who want to control entry to their highly paid profession.[11] To find a less self-serving answer, however, we have to turn back to the tension between the individual and the collective.

If the figure were to represent the truth of the matter, then a randomized control trial comparing bogus and qualified doctors' performance across a range of treatments would show that qualified doctors were, on average, a little better than bogus doctors. This means that so long as we have only population statistics to work with, we should always prescribe a qualified doctor. But this kind of population-average analysis obscures what happens in particular cases. Sometimes an experienced bogus doctor will do as good or better job than a qualified doctor. According to our analysis, the number of occasions when the bogus doctor will do at least as well as the qualified doctor is large: only rarely does a bogus doctor make a medical mistake which is serious enough to cause real trouble.

The logic of the argument is analogous to the one we used in chapter 1, where we compared the treatment of broken bones with randomized control trials. Population averages should be used only when there is a shortage of information about the causal chains pertaining to the individual patient (or patient-doctor interaction). With broken bones we do have enough information to forget about the population averages; with bogus doctors we usually do not.[12] Therefore, since, in the main, we can work only with averages, we are right to prefer qualified to unqualified doctors.

A second reason, also turning on the individual-collective axis, is that if it were accepted that anyone could walk in off the street and work effectively without medical training, then it would be hard to maintain that medical science makes more than a marginal input to health. If this were allowed to be recognized, medicine would revert to a folk practice in the long term.

We can see, then, that short-term and individual considerations point one way—to the cost-effectiveness of the benefits that bogus doctors can deliver—while long-term and collective considerations point the other way: on average bogus doctors are not so good as qualified ones, and the very idea of the bogus doctor is in tension with the idea of medical science as the predominant approach to health in our society.

The argument that the true problem is not the lack of skill of the medically unqualified is reinforced by the presence of other unqualified groups at or near the frontiers of medicine. At times of major accidents and natural catastrophes, on battlefields, and in cases of individual emergency (see chapter 6 on cardiopulmonary resuscitation), the boundary between the trained and untrained becomes blurred. Furthermore, nowadays the circle is being squared by the recognition of the medical efficacy of less qualified groups. Nurses are now allowed to be seen taking on increasing responsibility, while the new category of "paramedic" recognizes how much can be done in the absence of a complete medical knowledge. So long as such groups are officially partitioned off they can efficiently take on more and more of the role of doctors (as nurses have done without recognition for decades), without putting into question the long-term technical future of the profession. Society, while it now recognizes the paramedic, is not quite ready to give official status to the experienced bogus medic because we do not want any form of cheating in our lives. Coolly considered, however, the principle is to do with maintaining trust in socially defined roles rather than safeguarding the effectiveness of medical care.

Conclusion

Thinking about bogus doctors yields several lessons, some unsurprising, some not so obvious. To start with the unsurprising, bogus doctors reveal how much uncertainty and variation there is in medical practice. It is the large variation *between* the medical practices of different countries which makes it easier for bogus doctors to survive those tricky first few months of training on the job; at least some mistakes can be seen as stemming from difference of approach in the different countries in which the novices have been trained. It is the large variation in what can count as an acceptable treatment *within* any one national medical regime that allows bogus doctors to survive even while they are making what, in retrospect, seem like bizarre judgments (such as the Selsun prescription).

A less obvious lesson, but one of equal importance in terms of what we are exploring here, is again learned from the early period of a bogus doctor's career before experience has been acquired. This is that the surrounding team forgive mistakes because they assume that even a doctor who has passed successfully through the rigors of medical school will be an ignoramus when it comes to medicine as it is practiced. The bogus doctor case shows us how relatively unimportant book-learning is when it comes to understanding illness. This should give us pause for thought when we are trying to gather enough information from books and other written sources to challenge the medical profession.

The final lesson confirms the central argument of this book: to understand the way we respond to bogus doctors we must carefully separate the way we think about populations from the way we think about individuals and individual treatments.

3 Tonsils *Diagnosing and Dealing with Uncertainty*

Many people's first and only encounter with the surgeon's knife is the removal of their tonsils (and/or adenoids). Tonsillectomies became one of the first standardized operations, the hallmark of the new production-line surgery that emerged in the early part of the twentieth century. Tonsillectomies were thought to reduce sore throats and stem the sometimes deadly infections that accompanied them. And throughout the century tonsils were removed in vast numbers.

Sore throats have not gone away and tonsils are still being removed in large numbers today. In the UK, on average 80,000 tonsillectomies per year are carried out, mostly performed on small children. In the United States in 1996, the most recent year for which national data are available, 287,000 children less than fifteen years of age underwent tonsillectomy with or without adenoidectomy.

Tonsillectomies are not glamorous and are rarely newsworthy. There is the occasional lawsuit for medical incompetence when a child dies as a result of the operation. But there is usually no life-or-death drama associated with tonsil removal. No one is hoping for or waiting for a tonsil transplant. Tonsils seem all but useless. Recently tonsils, or rather their removal, have acquired a new significance in the UK. It turns out that the many thousands of tonsils removed every year are a good guide to the presence of variant Creutzfeldt-Jakob disease, better known as the human form of mad cow disease. A national archive for tonsil tissues was started in the UK in 2003 with plans to collect 100,000 pairs of tonsils from tonsillectomies. By testing for the strangely folded proteins known as prions, the discarded tonsils can be made to bear witness to the spread of this deadly disease.

The effectiveness of tonsillectomy is controversial and has long been debated among pediatricians. Over the years doctors have become more skeptical as to the supposed benefits of this operation. With the development of antibiotics, respiratory infections can be treated and managed by less drastic methods. The operation, although in decline, is still, however, the most common on children in the United States. The debate over tonsil removal continues among experts but does not elicit much wider interest, especially if we compare tonsils to other sorts of medical interventions. For instance, the steady rise in caesarian sections and decline in vaginal births raise recurring concerns about whether childbirth has become overmedicalized, perhaps through the increased pressure from lawsuits or from the doctor's timetable replacing the mother's timetable. It is the very lack of public debate, the lack of media attention, which attracts us to the case of tonsillectomy. What we want to do with this chapter is to examine the uncertainties surrounding this most routine of diagnoses and operations.

Medical uncertainty, as we said in the introduction, is scarcely news. We never have perfect knowledge when diagnosing and predicting the course of a disease and prescribing an effective remedy. Modern diagnostic tools may help, but in some cases they introduce new uncertainties. The new diagnostic tools of gene testing and mammography for breast cancer are typical. Such tests are notoriously unreliable guides to action—just because there is a statistical chance of a cancer forming doesn't mean that it will form. Should one take early drastic action or wait? The endemic uncertainties in the predictions for individuals, not to mention the difficult choices, are magnified by the fallible process of testing itself and the interpretation of the results. The false positive is never far away. Biotech companies who aggressively market their tests, public health lobbyists, feminist advocates, and patient activist groups reside on all sides of the debate.

Returning to tonsils, the need for their removal and the effectiveness of their removal, are, as we shall see, also accompanied by uncertainties. But in this case we will encounter the uncertainties in the raw, away from the arena of commercial and media interest and politics; we encounter the limitations to our knowledge and skills as the doctors and patients themselves experience them, in the routine work, the day-to-day life of medicine. Sore throats, small children, and those strange, sometimes painful, protuberances in the throat will be our guides.

We have divided the material into two parts. First we will look specifically at the case of tonsillectomies. Then we will explore the different sorts of expertise that patients and doctors bring to a medical consultation. By mapping out the different forms of expertise we hope to show why routine medical consultations can produce uncertain outcomes. Last, we return to tonsillectomies and ask what lessons there are to be learned.

What Do Tonsils Do?

Tonsils are small glands located on either side of the throat. They, and the closely related adenoids (masses of not normally visible tissue located at the rear of the nasal cavity), are part of our immune systems. These fleshy glands form part of an early warning system to help fight infection. Their location near the entrances to our breathing passages makes possible the early detection of viruses and bacteria which might be absorbed into our bodies from the air we breathe. When bacteria and viruses come into contact with our tonsils and adenoids, our immune system is triggered to produce antibodies to help fight off infection.

Tonsils and adenoids are thought to play an active part in fighting off disease in young children (under three years of age). For the first six months of the child's life they may even be crucial. It is less clear what purpose they serve for older children or adults. Our bodies, like safety critical systems in aircraft, have built in redundancy. Other parts of our body, such as circulating T Cells in the blood, also effectively trigger the immune system, and the adenoids even whither away naturally in teenage children. Many people (including one of the authors of this book) have had their adenoids and/or tonsils removed in childhood and appear to be none the worse for it.

But tonsils can cause trouble. Many of us experience some swelling of our tonsils when we have colds or sore throats; sometimes this swelling becomes painful and the tonsils are inflamed with white blotches—a condition known as tonsillitis. In some children this is a recurring condition, and it can lead to chronic difficulties with breathing and swallowing accompanied by infections which spread throughout the ear, nose, throat, and lungs. Strep infections are particularly dangerous because untreated they can lead to life-threatening diseases like rheumatic fever. Swollen tonsils can become dangerous as obstructions and they can prevent normal sleep. The removal of tonsils means there is one less organ to get sore and swollen and less chance of the associated problems. The logic is simple. Some children will benefit if diseased, painful, and swollen tonsils that serve no obvious purpose are removed. Cutting them out is easy and there are no apparent ill effects to the body from their loss.

A Brief History of Tonsil Removal

Tonsils have been removed since at least the first century AD. A technique for the operation was described by the famous Roman physician Celsus. Early procedures were risky and painful, involving either direct cutting with a knife or strangling the tonsils with a soft wire—a procedure which could take up to twelve hours, during which the patient sat drooling in pain and unable to swallow. The invention of the tonsilotome by the appropriately named Philadelphian physician, Philip Physick, in 1832 made the surgery much less painful. Physick adapted an existing instrument, the uvulotome, which had been used since the seventeenth century in Norway to remove the uvula, the small piece of tissue hanging in the back of the throat. The tonsilotome comprises an enlarged ring with a piece of waxed linen behind to support and hold the tonsil in place to ensure a clean cut by a retractable blade as in a guillotine. Most operations today are either carried out by dissection or by one of the many forms of guillotine tonsillectomy.

It wasn't until the early twentieth century that the practice of removing tonsils and adenoids began in earnest and paralleled the growth of surgery itself. Tonsillectomy rapidly became the most common form of surgery. Historians of medicine have studied the sorts of operations carried out in early hospitals. For instance, in 1895 at the Pennsylvania Hospital in America (founded in 1751 and one of the oldest, if not *the* oldest, hospitals in the United States) the most frequently performed operation was excision of the cervical adenitis (that is, removal of inflamed lymph glands in the neck), and it was performed twenty five times. Three decades later, in 1925, the most common operations at the same hospital were tonsillectomies and/or adenoidectomies, and over one thousand were carried out. (The next most numerous operation was the appendectomy, of which there were 234 that year.) The first series of the *Surgeon General's Index Catalog* published in 1893 had barely three pages of references on surgery of the tonsils. By the time it was updated in the 1913 second series, there were eighteen pages on surgery of the tonsils alone.[1]

Adenoidectomy has a more recent history. Wilhelm Meyer of Copenhagen in the latter half of the nineteenth century suggested that adenoid vegetations were responsible for nasal symptoms and impaired hearing. Opinions have differed over time as to whether surgery should consist of just tonsillectomy or adenoidectomy or the combined procedure. Removal of tonsils is usually considered efficacious for throat conditions and removal of adenoids for diseases of the middle ear. In practice, surgeons have often favored doing both operations together to take advantage of the patient's hospitalization and anesthesia.

The reason so many tonsils and adenoids were being removed in the latter half of the nineteenth century was the then popular "focal theory of infection." This idea was part of the microbiological revolution brought on by the germ theory of disease in the 1870s coupled with new technologies like x-rays (invented in 1895), which enabled doctors to look inside the body at areas previously hidden from view.[2] If microbes caused disease, then any location which harbored microbes was likely to be a source of diseases ranging from arthritis to nephritis (inflammation of the kidney). As it was noted at the time, the tonsils "form an ideal nest for the development of micro-organisms. There are warmth, moisture, decomposing secretions, and a harbor from the currents of air or friction of fluids that might otherwise dislodge them" (Howell, 60).

The increase in tonsillectomies has, of course, been accompanied by an increase of diagnoses recommending removal of the tonsils. Indeed, as tonsillectomies became more and more common it seemed that indications for removal of tonsils became practically coextensive with the existence of the organ itself. As one leading physician recalled in the 1930s, "Almost all children had diseased tonsils that were a menace to life and health." Diagnosis was based on two sorts of indications: physiological and pathological. Physiological indications were impairments of some type that could be linked to the tonsil, such as an "unpleasant quality to the child's voice" or severe earaches and impaired hearing or "certain failures of the body mechanism, leading to a lack of resistance or a general indefinite *below-par-ness*." Pathological indications included evidence of infection elsewhere or simply prophylaxis—because the tonsils were there they should be taken out. In other words, medical evidence alone gave grounds to justify removing the tonsils of almost any child.

The removal of tonsils in ever increasing numbers was itself a big spur to the development of hospital surgery. More children seemed to need the operation than could be accommodated at hospitals. A 1920 survey of children in New York City revealed that between 10 and 20 percent of children had either enlarged tonsils or defective breathing—sure signs that their tonsils needed to be removed. Some medical leaders at the time considered letting less-experienced practitioners, such as pediatric interns, remove tonsils in order to get more operations done. In the early part of the century no other operation was thought to be a more compelling demonstration of the power and success of surgery. As one historian notes, "No operation in surgery yielded a higher percentage of satisfactory results" (Howell, 61) It is worth noting also that no other operation produced a more satisfactory result in terms of benefit to surgeons' bank accounts. Surgeons could get rich on tonsils. Modern studies of group surgical practices, where prepayment insurance programs are not geared to the volume of surgery performed, show a lower frequency of tonsillectomies.

Epidemiological Puzzles

But just who were getting their tonsils removed? Epidemiologists noted puzzling trends: certain groups of children seemed more likely to lose their tonsils than others. One early study was of tonsillectomies carried out in the 1930s in the UK under the auspices of the Free School Health Service. Its author noted, "Comparisons of some of the rates in different areas in 1931 . . . revealed striking contrasts in areas apparently similarly circumstanced. Thus in that year the operation rate in Margate was eight times that in Ramsgate [two similar seaside resorts]; that of Enfield was six times that of Wood Green and four times that of Finchley [similar areas of London]; that of Bath five times that of [the neighboring] Bristol; that of Guildford four times that of [close-by] Reigate; that of Salisbury three times that of [near neighbour] Winchester" (quoted in Bloor, 44).

Studies of the class backgrounds of patients revealed intriguing findings; for example 83 percent of new entrants to Britain's leading fee-paying school, Eton, had in 1939 arrived without tonsils. Comparisons with other operations also revealed perplexing data. One study carried out in 1950 found that children in Canada who had undergone appendectomy were twice as likely to also undergo, or have undergone, tonsillectomy. Perhaps the most bizarre finding of all was that of Miller and his associates in 1960, who found that in their large sample of children in Newcastle, boys under four years of age were seven more times likely to have undergone tonsillectomy if they had been circumcised.

How were such startling variations to be explained? Studies started to focus upon the process by which it was decided whether the operation was really necessary. One classic study was carried out in the United States before the war and involved 1,000 New York schoolchildren. Sixty-one percent of these children had already lost their tonsils; the remaining 39 percent were assessed by a group of school doctors, who recommended that a further 45 percent should undergo the operation. The "rejected" children were then sent to a second group of doctors, who recommended surgery for 46 percent of them. The remaining children, now twice rejected for tonsillectomy, were then sent to a third group of doctors, who recommended surgery for 44 percent of them. At this point only sixty-five of the original 1,000 children had not been operated upon or had the operation recommended for them.

The lack of reproducibility of examination findings in adeno-tonsillectomy cases was found to occur not only *between* different doctors but also for the *same* doctor over time: in one British study nine color slides of children's tonsils were shown to a sample of forty-one ENT (ear, nose and throat) specialists, pediatricians, and GPs. Unbeknownst to the doctors, two slides were shown twice; it was found that the average ability of the doctors to arrive at the same assessments of the twice-shown slide was only slightly better than chance.

Medical Examination under the Microscope

It is clear from the sorts of studies we have discussed that different doctors reach startlingly different assessments as to the need for a tonsillectomy. One explanation might be that the diagnostic criteria which doctors use vary widely. This turns out not to be the case. The sorts of indicators which doctors refer to turn out to be rather standard in nature, although different authorities place different weights on these indicators. To find out what is leading to the different conclusions we turn to one detailed sociological study of how doctors actually make their diagnoses. This study was carried out by medical sociologist Michael Bloor in the UK in the 1970s. Bloor observed eleven different specialists at a number of ENT outpatient clinics held in hospitals in the UK. Within the UK such specialists work at one level removed from ordinary family doctors. The GP will typically see a child at his or her practice and then if necessary refer that child for consultation with an ENT specialist (usually at a local hospital). It is the ENT specialist who makes the final decision and who will carry out the surgery if it is necessary. The task for the specialist is to assess each case and decide whether a tonsillectomy is really appropriate.

It is important to realize that from the viewpoint of the specialist such assessments are routine. They examine children with sore throats day in and day out, year in and year out. The children always exhibit a number of symptoms within a fairly limited range of morbidity. The patients' complaints are familiar and the doctors follow familiar procedures of investigation that enable them to produce familiar prognoses indicating a familiar form of therapeutic intervention. In other words the diagnosis does not concern a life-threatening condition made under conditions of great duress, and the operation itself is a routine one. It is the very routine nature of the process which makes it so illuminating in terms of medical decision-making.

Bloor observed that all the specialists began the examination of the children by looking for three crucial clinical indicators for a tonsillectomy: (1) enlarged cervical glands (lymph glands which when enlarged can be felt on the surface of the neck below the ears), (2) pitted tonsils;, and (3) a flushed appearance to the anterior pillars which flank the tonsils in the pharynx. The specialists differed, however, over which of these signs they regarded as relevant to their assessments. In particular some doctors looked for a broad range of signs, while others looked for a narrow range. For instance, one doctor viewed the presence of any one of the three main signs as evidence of tonsillar infection and hence the need for removal. Another doctor's sole criterion was multiple enlarged cervical glands or two cervical glands so enlarged that they were clearly visible on the surface of the neck.

A more telling difference was not over which clinical signs to look for but the importance of these signs in relation to the particular patient's history. And here we encounter for the first time the full complexity of the task faced in medical examination. Producing a patient's history is not a trivial matter, yet much medical diagnosis and examination depends crucially upon it. As we noted in the introduction to this book, medicine as a profession achieved a breakthrough in the late nineteenth and early twentieth century when, through the use of postmortems and new medical technologies like the stethoscope, it was able to make diagnosis less dependent upon patients' own self-assessments of their various illnesses and ailments. Recommending a tonsillectomy, however, turns out to one of those cases which still depends both upon patients' self-reports (of sore throats, swollen tonsils, and the like) as well as direct clinical evidence visible to the doctor. Furthermore, in the cases Bloor observed, two different layers of the medical profession—the GP who first encountered the patient and the surgeon who had to do the operation, assessed the evidence. This, as we shall see, complicates matters even more.

Bloor found that different examiners adopted different strategies in order to deal with patients' histories. One doctor in Bloor's sample considered the appearance of two out of three main clinical signs as crucial, and these signs were taken to be the *sole basis* for diagnosing the need for a tonsillectomy. In other words the patient's history was unimportant. This doctor found that the majority of his cases were those with two or more clinical signs: "I would think that probably eighty percent of those that I see—you know referred for tonsillectomy—four out of five fall into this category" (Bloor, 48). If there were no direct clinical signs of tonsillar infection, then an immediate operation was not needed in this doctor's view.

But another doctor adopted almost exactly the opposite strategy. This doctor attached no significance at all to examination findings relating to the tonsils or cervical glands. On some occasions (for example, if a child became upset or frightened in the clinic) he would even forgo any attempt to examine the child and make his decision purely on the basis of the history, in most cases supplied by a parent. Another doctor similarly emphasized the importance of a history of sore throats: "I think in almost every instance it's the history rather than the examination. Somebody is supposed to have said once that the only point in looking at the child's throat is to make sure the tonsils are *still* there, that nobody else was there before you! This is an exaggeration I'm sure, but it puts the point over" (Bloor, 49).

History, of course, is seldom neutral. Even the telling of a simple piece of history about sore throats can involve many different sorts of assumptions. For instance, how much detail should the doctor search for in recovering the history from the patient? Bloor found that one doctor was content simply to seek affirmations of his questions to the parent as to whether the child had been extensively troubled by sore throats:

Doctor: Is he having an awful lot of trouble with his tonsils?
Mother: Yes.

On the other hand, some of the doctors interrogated the parents extensively as to the duration of the symptoms and as to the frequency or reoccurrence, e.g.:

Doctor: She's getting sore throats?
Mother: Yes.
Doctor: How long has this been going on?
Mother: For two years in the winter.
Doctor: How many times in a year do you have to get the doctor's help?
Mother: Three or four times.

Some of the doctors chose to pursue a wider range of symptoms as relevant for this history. In the case below the severity of the attacks and the presence of any aural symptoms are both raised.

[A ten-year-old girl was referred with a GP's history of "recurrent" sore throats and earache. The GP mentioned enlarged and "unhealthy" tonsils and wondered if tonsillectomy was indicated.]

Doctor:	How often have you been having sore throats?
Mother:	Quite recently it's been a lot.
Doctor:	How long do they last?
Mother:	A week at a time.
Doctor:	This been going on for what? the last 2–3 years?
Mother:	Yes.
Doctor:	How many attacks a year?
Mother:	3 or 4.
Doctor:	She needs to get penicillin for them?
Mother:	Yes.
Doctor:	Losing a lot of school?
Mother:	A fair amount.
Doctor:	Any ear trouble?
Mother:	Aye, she had earache the last time.
Doctor:	Is it usually in the winter she has trouble?
Mother:	Yes.
Doctor	(examining the child):The ears look alright. How old is she? About 9?
Mother:	10.

[Dr. found infected tonsils, large adenoids and negative findings for the nose and ears. He listed the child for T's and A's. (Tonsillectomy and Adenoidectomy)]

By requesting more specific information, some of the doctors were able to make an assessment of symptoms that was independent of the parent's and GP's assessments. Thus, by inquiring after frequency of attacks, the doctor was able to come up with an assessment which, although dependent upon parental reportage, could differ substantially from the parent's own assessment of the condition. Furthermore, by looking for indications such as whether antibiotics have or have not been administered in the past, the specialist can again find a criterion independent of the parent's own assessment—although in this case it makes the specialist more dependent upon the GP's earlier assessment of the severity of the condition. In one case Bloor asked a specialist after a consultation whether a particular child had tonsillitis. The spe-

cialist remarked, "Yes I think so. The GP wouldn't put him on antibiotics just because he's got a cold" (Bloor, 51).

The very fact that children were there as a result of referrals
thus led to different emphases placed on different parts of the
patient history. For one doctor the fact of referral itself was an
indication of a history of recurrent infection, and his truncated
history-taking was simply designed to confirm this, along with
examination evidence. On the other hand, some of the doctors
sought out much more elaborate pieces of evidence, never taking
referral itself as indicative of much.

We can now start to see why different doctors reach different
conclusions. It is not that some doctors are poorer at diagnosis
than others (although this indeed may be the case) or that doctors
are looking at different indicators and symptoms. The doctors
seem to agree on the general sorts of symptoms and appropriate
indications for surgery in the abstract, but the particular ways the
doctors use and routinize these general conditions can vary dramatically. The set of symptoms that everyone agrees are appropriate indicators for removal of the tonsils are interpreted in different
ways in the daily practices of specialists.

There is one last complication in these examinations. They deal
with children. Often doctors appeared to be content, as in the
above cases, to rely upon parents' accounts as to what symptoms
appeared, when they appeared, the course they took, what treatments had been administered, and so on. This adds a layer of
complexity because it requires the parent to be a reliable witness
and interpreter of the child's symptoms. Doctors, of course, even
when dealing with adults, have no direct access to the patient's
own subjective state of mind—they must rely upon a mixture of
self-reporting and direct observation. For example, a patient can
say, "This pain is killing me," and the doctor can verify this by
noticing that the patient winces when she sits down in the doctor's
office. But drawing inferences from these two sources of informa-

tion becomes even more of an art form when dealing with children, who may have a limited vocabulary and a less well-developed sense of what is meant by a faithful description. For instance, the four-year-old daughter of one of the authors of this book once complained of a pain somewhere in her leg but refused to say where it specifically was. A pediatric specialist proceeded to examine her by making her walk in various ways around her office. The art was to get the leg to "speak for itself" as it were, as to where the pain was.

Bloor found the age of the child was often crucial to the different routines the doctors adopted. Thus one specialist who usually relied on the parent's assessment as to the severity and duration of attacks would ask much more specific questions for very young children. Others, when dealing with very young children, would try to assess the impact of the infection on the child's general health. In other words, as might be expected for each of the specialists, symptoms and signs took on different meanings when manifested in children of different ages. Judgments also varied as to what counted as a "younger child." Two specialists used special routines for two- and three-year-olds. Another employed very different decision rules for children under seven. One specialist had the most elaborate age-related scheme of all, distinguishing three age groups: a group aged three or younger, for whom the criteria for admission for surgery were most restrictive; a group aged four to six, for whom his criteria were less restrictive; and a group aged seven and above, for whom his criteria were least restrictive.

We stress again that all the specialists in this study were competent, highly experienced examiners. The different judgments they reached did not seem to result from differences in training. Indeed, they all agreed upon the sorts of criteria that were relevant. Rather the uncertainties seemed to be endemic to the very process of medical diagnosis itself.

Diagnosing Diagnosis

To try to understand why there is so much uncertainty in medical examination and diagnosis we shall leave tonsils behind for the moment and look at visits to the doctor in terms of the various sorts of skills and expertise involved. All diagnosis involves weighing a variety of pieces of clinical information along with a patient history that has been produced with input from the patient, and sometimes other patient representatives like parents, and other experts, such as the referring GP. Producing the final diagnosis therefore depends upon many different sorts of skills and expertise.

Working outward from the perspective of the patient, we first encounter the skill or expertise of the patients themselves in recognizing, classifying, and diagnosing their own symptoms. Self-diagnosis usually precedes any visit to the doctor. Here elementary skills of observation and memory are involved. Some people are better than others at observing and monitoring their own bodies and noticing their state of health over a period of time. Part of the skill in moving from observation to self-diagnosis is to know when something is wrong; when a symptom is abnormal and serious versus normal and trivial. Children have to learn this skill, and part of being a good parent is learning to read the subtle clues which tell you that your child is really sick (and sharing that knowledge with your children so they can learn for themselves how to do it). Self-diagnosis is complicated by individual history and variability and by the fact that many nonspecific symptoms are routine—upset stomachs can just be part of life, as can low-grade aches and pains like headaches. Being familiar with your own history and variability well enough to know when something is wrong seems to be a key element in making a satisfactory self-diagnosis.

One element of the skill in self-diagnosis is the problem of assessing whether physical symptoms are real or whether they have a psychosomatic element. This distinction itself is of course not always clear-cut, and the recognition that there is a psychosomatic element may still mean the symptoms warrant attention. I may feel generally lousy today and notice I have muscle ache, but if I feel low every day, my muscle ache may seem to take on a life of its own. Conversely, I may have genuine pain that leads to me being depressed, which can further accentuate the pain. As we pointed out in chapter 1, the mind-body interaction is a vast area of uncertainty for medical science—it is one of the few areas where individual patients familiar with their own moods and ups and downs can supplant the necessarily less immediate knowledge of the doctor. There is, however, the danger (which we point to and discuss in more detail in chapter 5 on yuppie flu), that patients may misdiagnose these psychosomatic symptoms as part of a new and questionable illness. While individuals can skillfully monitor whether their symptoms follow a pattern and whether or not they fit a diagnostic category, that disease category itself can be established only by medical science, such as with careful epidemiological studies—something no individual patient is in a position to definitively determine. The extent to which a patient can go beyond self-diagnosis and become, as it were, a medical scientist is one we return to in chapter 5.

Lastly, it is worth pointing out that self-diagnosis is a process which involves using many of the same interpretive skills that doctors themselves use in diagnosis. For instance, self-diagnosis can involve observing and interpreting the outputs from medical instruments. Patients increasingly rely on instruments as part of self-diagnosis—whether the simple thermometer of old, the blood pressure meter, the blood sugar indicator, the heart monitor, or the peak flow meter. The technology of medicine is no longer restricted to the professional, and patients are regaining some of the power once lost. The patient, like the doctor, has to render readings from these instruments into some sort of objective, reportable data. Similarly patients have to monitor the effects of various medical interventions, ranging from drugs to pacemakers, **77** upon their own bodies. Symptoms have to be noticed, categorized, and classified and then interpreted within the context of some disease or chronic condition. Patients may become good at identifying the causal elements of their disease condition. This is particularly true of chronic conditions like asthma, where patients routinely learn what sorts of allergens to avoid. How competently the patient carries out a self-diagnosis thus will depend on a myriad of factors, including her own observational skills, her training (some patients are taught by doctors, nurses and health visitors to use instruments and make interventions at home) and her exposure to medical knowledge and practice (and here, having access to books about medicine, self-help manuals, and the Internet can make a significant difference).

The ability to self-diagnose probably varies from culture to culture. For instance, we are struck by how in the UK health system, with little choice over doctors and a general ethos of trusting doctors, there is a low medical literacy (even among the well-educated classes) compared to the United States, where people choose doctors more freely and often seem to display familiarity with a richer medical vocabulary for describing their various ailments and courses of treatment.

A doctor who is him- or herself a patient will presumably be fairly good at self-diagnosis; likewise, someone who has visited the doctor fifty times with the same set of symptoms will presumably be better than someone going for the first time; and someone who watches TV medical dramas will presumably be better than someone who watches TV game shows. In all cases we can get more skilled over time—whether in taking a temperature or monitoring blood-sugar level or giving a shot or understanding the course of our own ailments. In many senses the patient quickly becomes the expert on her own illness and its course, but this localized expertise born of long experience should not be confused with generalized expertise in medicine.

"Self-diagnosis" is in many cases a misnomer. Self-diagnosis can and frequently does involve a collaborative effort—a kind of pooling of widespread expertise. This is most obvious in the case of children and old people who depend upon others to recognize their symptoms for them. But many of us discuss our symptoms and "what it is we've got" with family, friends, and colleagues. Internet and online chat rooms and listservs can provide a further source of collaborative effort—especially for unusual and new ailments.

When we visit the doctor, the expertise at self-diagnosis has to be accompanied by another form of expertise, in which the doctor also participates. This is the skill in *translating* your own self-diagnosis, your own symptoms, and your own history of the illness into something that the doctor can use as input into his or her diagnosis. Obviously a key element here is the skill of the doctor in interacting with you, the patient. Medical sociologists have analyzed the doctor-patient interaction in great detail. It is a complicated business which can involve role play, language skills, class background, ability to elicit empathy, sympathy, and rapport (what is often called bedside manner), and subtle interactional factors such as how the body should be presented in medical examination, the appropriate gaze of the patient and doctor, and so on.

Treating this interaction as a form of expertise that the patient and doctor both possess involves asking new questions. For example, as a patient you have a choice: do you provide your analysis of your own symptoms, or just describe them, or do a bit of both? How do you deal with the problem of the theory-ladenness of observations—do you describe the rash as a rash or as a scarlatina (the allergic rash produced during scarlet fever), do you mention the strawberry tongue you have observed or leave it to the doctor to find it himself? Once you have a self-diagnosis, you know what to look for, and this may color an impartial examination. But on the other hand, if you are certain that you know what you have (you may, of course, be wrong), your goal may be to do just this—that is, persuade the doctor as quickly as possible to agree with you and get out of there with the "correct" diagnosis and prescription. How do you then describe and present symptoms in a compelling way so as to make the doctor take notice? We've all experienced that strange, reverse psychosomatic effect whereby the stress of seeing the doctor actually makes you feel better (this effect is well known in other areas like teaching, acting, and TV interviewing, where the stress of performance gives more control over bodily functions than is normal—e.g., actors rarely sneeze, break wind, or have other uncontrollable bodily emissions when on stage). Memory, rhetoric, and the ability to interact with humor can all affect the outcome, and there is always the second-order interactional effect: that the doctor may be trained to recognize and accommodate to your lack of skill in presenting symptoms and diagnosis and that by making a rhetorically forceful case—hamming it up, as it were—you may actually make a less compelling case. It is not that medical diagnosis is somehow independent of such factors—of course doctors can be trained to recognize and if need be minimize such factors—but the patient history is a joint accomplishment; the doctor too must display interpersonal expertise in eliciting the patient's account and part of his or her expertise must be in assessing the veracity, consistency, and plausibility of your account. As we saw in the tonsillectomy case, independent kinds of evidence such as referrals from other GPs, evidence of drugs prescribed, and so on also form part of the information available and can be used as benchmarks to assess the patient account.[3]

Conclusion: A Spectrum of Diagnostic Uncertainty

Eventually a diagnosis or recommendation will be made. The tonsillectomy case shows how it is possible for doctors drawing upon similar sorts of indicators to reach different decisions. If all the different sorts of expertise involved were standardized, then perhaps we could expect more uniformity in the outcomes. Unfortunately this is not so. Perhaps the most standardized form of expertise comes from medical training. Doctors are trained in similar ways (although even here forms of training may vary between medical schools and in different national contexts—for instance, in the UK doctors are trained to use the stethoscope for monitoring subtleties in heart conditions; in the United States they are not). The role of such common medical training was apparent in the tonsillectomy case—all the UK doctors agreed upon the sorts of symptoms that were relevant to the tonsillectomy decision. Even possessing a common training, however, doctors have different amounts of experience. More crucially, in cases like tonsillectomy, where a chronic condition is evaluated over a period of time, there must always be input from the patient, and patients are not trained in a standard way. Their expertise at self-monitoring and diagnosis can vary enormously.

The uncertainty around diagnosis varies not only from case to case but more importantly from disease to disease. We can now clarify how we should think about uncertainties in diagnosis in terms of our theme of the individual versus the collectivity. Some diseases are well enough understood for an individual diagnosis to be highly reliable (although the particular diagnosis may vary from case to case). As we mentioned in chapter 1, at one extreme are ailments like broken limbs, where the cause is well understood and where it would be ludicrous to apply the statistical thinking such as produced by RCTs. Diagnosing a need for tonsillectomy is toward the other end of the spectrum. It is more like seeking an answer to whether diet is correlated with heart disease or cancer. What we ideally need is statistical evidence such as that provided by an epidemiological study. Tonsillectomies are not, of course, diseases like heart disease or cancer. A tonsillectomy is a procedure and hence more like a treatment. As we have reported earlier in this chapter, epidemiological evidence has shown how tonsillectomies are correlated with what we usually take to be irrelevant sociocultural variables such as education, circumcision, and nationality. This indirectly throws into question their effectiveness. But there is a more direct statistical way to assess the effectiveness of tonsillectomy; that is, conduct an RCT where a population of sick children are given the treatment "tonsillectomy" and are then compared to a matched group given no treatment at all (it is hard to imagine how this could be done in the strict way of administering a placebo to the control group, because one could not conduct fake surgery on a large population). Although such evidence would not tell us directly whether tonsillectomy was effective in any individual case (because it is possible that people get better for reasons completely unconnected to the tonsillectomy), it could discriminate for the group as a whole whether tonsillectomy is a better or worse intervention. In recent years such trials have in fact been conducted on a group of children in Pittsburgh. One study used children with recurrent severe sore throats (the children met far stricter criteria for being prescribed a tonsillectomy than the standard criteria). The results showed that although children in the "no-treatment" limb often got better of their own accord, tonsillectomy produced more benefit in terms of less reoccurrence of sore throats. A follow-up study, however (reported in 2002), on children who met less stringent criteria for prescribing a tonsillectomy (but still more stringent than the standard), found only marginal benefits from the surgery. As the authors concluded, "the modest benefit conferred by tonsillectomy and adenotonsillectomy in children moderately affected with recurrent throat infections seems not to justify the inherent risks, morbidity and cost of the operations."[4] If we can accept the findings of this study, they indicate that for recurrent, nonsevere sore throats, tonsillectomy and adenotonsillectomy should not be recommended.[5]

Uncertainty in medicine is endemic, but it is crucial to know where you are on the spectrum of uncertainty that stretches from cases where the intervention at the level of the individual is well understood (broken limb repair) to much more uncertain interventions where we have only statistical data, such as most cases of surgical removals of cancers. It is rare to seek a second opinion for a broken limb (unless the fracture is complicated and requires a complicated procedure), but if you have been diagnosed with breast cancer you would do well to seek a second opinion. The strange thing about the case of tonsillectomy is that we tend to treat it as being more toward the broken limb end of the spectrum. It is unusual, for instance, to seek a second opinion about a tonsillectomy. Perhaps the routine nature of the operation and its apparent low cost and safety are deceptive. The perception seems to be that having a tonsillectomy is not a big deal. But it can be a big deal, especially if something goes wrong, and occasionally it can (hemorrhage is the biggest problem). Any operation involving anesthetic (as a tonsillectomy does) runs a risk. There is about a 1 in 14,000 chance (unadjusted for age) that an adverse reaction to the anesthetic will produce death. Cases of death stemming from the operation itself are rare, but do happen from time to time. Unfortunately, accurate mortality statistics in large patient populations for this operation are not available. Case fatality rates have been reported since 1970 that vary between 1 in 1,000 and 1 in 27,000.[6] Although the risk from death seems small, it is on some counts about the same risk as of a crib death. Crib death attracts media attention and money. The suddenness and unexpectedness no doubt play a role, and unquestionably to lose a baby in this way is one of the cruelest and most devastating events we can experience. Deaths from a tonsillectomy on the other hand are treated as banal. There is no scandal or public outcry about death from this supposedly routine operation. But, then, death from car accidents is routine—a staggering 115 deaths every day in the United States—yet we worry far more about plane crashes, terrorists, and vaccinations. This chapter reminds us that it is routine uncertainty and routine death that are routinely overlooked.

Alternative Medicine *The Cases of Vitamin C and Cancer*

An air-conditioned modern tourist bus is about to leave the conference center of the Hungarian Institute for Environmental Research at Tihany on Lake Balaton. On board is one of the authors (Pinch), along with the other participants from a four-day international conference on "Hermeneutics and Science." They are heading back to Budapest and their international flight connections. There is sudden commotion as the conference organizer, a young Hungarian woman who had earlier presented a paper on the philosophy of mathematics, steps off the bus with the wife of a Japanese participant. She asks for a cigarette lighter. The Japanese woman attaches tiny incense candles mounted on small metal pads to the neck and hands of the Hungarian woman. The candles are lit and burn down. Pinch steps off the bus to see what on earth is going on.

What is going on is alternative medicine, Japanese style. The Hungarian woman has a headache, and the Japanese woman has offered to cure it with her portable acupuncture kit (see figure 4). His curiosity piqued, Pinch asks the Hungarian lady why she didn't simply take an aspirin. She tells him she wanted to "try

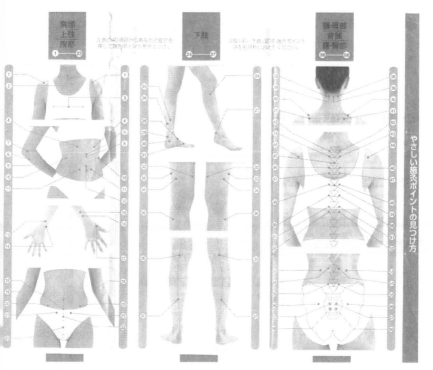

Fig. 4. Japanese portable acupuncture kit instructions

something different." Later Pinch asked her by e-mail whether the treatment had worked; she replied that it had worked "for a while."

The encounter described above, with its uncertain outcome, is by no means atypical. Today, with international travel, globalization, and more and more choice over health care, alternatives (sometimes known as "complementary medicine") that were once scarce and exotic are now only a click away. There is an enormous range of alternative therapies, including osteopathy, acupuncture, aromatherapy, Alexander technique, homeopathy, massage, shiatsu, iridology, chiropractics, herbalism, meditation, holistic reflexology, kinesiology, hypnosis, and all sorts of vitamin remedies. Many of these treatments parallel standard medicine in that they

85

offer drugs available in over-the-counter forms at pharmacies, health food stores, or via the Internet. A visit to the local health food store reveals a bewildering array of remedies ranging from Saint-John's Wort for depression to homeopathic pills in which, according to the standard scientific view, all active ingredients have been diluted to extinction.

Folk and quack remedies, of course, have been part of the healing enterprise throughout history. Indeed, for many of the afflicted that is all there was (and still is, for many parts of the globe). Some of the folk remedies still prevail but have now been brought under the umbrella of orthodox medicine. One of the earliest diagnosed diseases was the chronic form of arthritis known as gout (recognized since at least the time of the Greeks). The drug derived from the autumn crocus, colchicine, used for centuries as one of the few remedies for gout, is still the frontline drug for gout sufferers (interestingly, its effectiveness remains unexplained by modern medical science). The inexorable rise of modern medical science meant that by the middle of the twentieth century folk cures and remedies had either been coopted or displaced to the margins as "alternative medicine."

The recent extraordinary growth in alternative medicine seems to be a by-product of the sixties counterculture with its focus on Eastern religions, the whole body as well as the whole earth, and its suspicion of capitalism and its products. Before the 1960s, alternative medicine in the United States was often associated with the right wing, and indeed many of the leading practitioners in the United States in the 1930s were right-wing, anti-Semitic populists.[1] They accused the AMA (American Medical Association), which led the attack upon them, of being under the sway of Communists and Jews. In the UK at the time of the establishment of the National Health Service in 1948, alternative medicine had almost dwindled away, leaving, according to medical historian Roy Porter, a "small fringe of herbalists, mediums, faith-healers and spiritualists" (Porter, *Greatest Benefit*, 688). Porter notes, however, that by 1981 there were estimated to be actually *more* alternative practitioners (30,373) than general practitioners (30,180). In the United States the growth seems to have come a little later, but by 1999 the Stanford Center for Research in Disease Prevention was reporting that it found that 69 percent of respondents in a survey used some form of alternative medicine. According to the AMA, between 1990 and 1997 there was a 47 percent increase in the total number of visits to alternative practitioners. In the United States the strange alliance between right-wingers and liberals can be seen at trade shows for alternative medicines, which regularly attract members of antigovernment militias as well as new age enthusiasts. The National Institutes of Health, under pressure from Congress (including influential Republican and Democratic congressmen), established the Office of Alternative Medicine, renamed in 1998 as the National Center for Complementary and Alternative Medicine (NCCAM). With a budget of $90 million a year, the NCCAM encourages researchers to evaluate unorthodox treatments neglected by mainstream medicine, including bee-pollen supplements, electrochemical currents, intercessory prayer, and unorthodox cancer therapies.

Mainstream medical organizations such as the American Medical Association and its sister organization the British Medical Association have traditionally led the fight against alternative medicine. In Britain as late as the 1980s the BMA's medical ethics handbook threatened that doctors who had dealings with osteopaths and similar healers could expect to be subject to disciplinary proceedings. But by the 1990s the BMA, partly under the urging of its president, the Prince of Wales (the British royal family have long been proponents of homeopathy), had adopted a more conciliatory stance. Indeed, today many complementary treatments are available under the NHS (although GPs retain clinical control), and two in five GPs in 1998 referred patients to complementary therapists. The growth has been equally great in other European countries such as the Netherlands and France. In 1990 Americans made 425 million visits to unconventional healers compared with 388 million to primary care physicians.

Patient demand is growing. Desperately ill people, for whom orthodox medicine can do no more, will seek relief anywhere. It is the recalcitrant chronic illnesses (such as asthma) and low-grade ailments (such as backache), however, that provide the "bread and butter" for these alternative practitioners. When faced with a non-life-threatening but still painful and debilitating condition, the possible benefits of going "quack" often seem to exceed the risks. Why not give it a shot when orthodox medicine appears to offer so little, at best alleviating symptoms rather than treating the underlying cause?

There is no doubt that many readers of this book, like its authors, will have dabbled with herbal remedies, chiropractic, homeopathy, acupuncture, and the like. Such alternatives are today often included in health insurance policies, and it is routine in some national medical contexts (e.g., in Switzerland) to consult a homeopathic doctor as well as an "ordinary" doctor. Patients pass on information to each other—"the best chiropractor in town" or the one who "made my back worse"; the name of that Chinese herbalist in New York City who "cured my son's skin disease," and so on. But how effective is alternative medicine?

Evaluating the success of any form of medical intervention is extremely difficult. As we shall see in the case of cardiopulmonary resuscitation (CPR) (chapter 6), even the success of widely used orthodox techniques is hard to assess. Techniques based upon completely different cosmologies and practices are likely to be still harder to evaluate. Furthermore, alternative practitioners often refuse to accept the reductionist disease classifications of orthodox medicine. For example, the Navaho have a healing ceremony called the "Night way" for treating together what we would count as a diverse collection of diseases of the head, including head-aches, poor eyesight, and bad dreams. Alternative treatments may sometimes include elements that could belong with the orthodox repertoire (such as the "sweats" included in the Night way), and conventional notions of statistics may not apply to treatments aimed at the "whole person" rather than a well-defined "symptom." Adding to the difficulty of evaluating such treatments, patient records are often frustratingly incomplete at alternative clinics.

The most scientifically rigorous means we have for assessing the claims for the majority of medical interventions are randomized clinical trials (RCTs); they are legally mandated in many countries. For the reasons given above they are rarely applied to alternative medicine. This is a problem which Dr. Stephen Straus, who was appointed director of the NCCAM in 2000, set out to address. Straus announced his intention to apply "the same designs that are used in definitive studies of conventional medicine" to alternative medicine, including the use of random double-blinded trials. "Since gaining acceptance for alternative medicine is the objective," he said, "it is important to convince physicians, scientists, and pharmacologists . . . that studies have been done well and that the answers are as definitive as possible" (quoted in Juhnke, 152). In this chapter we shall focus on one example of an alternative cure evaluated in just this painstaking way—the claim that large doses of vitamin C can cure cancer.

Linus Pauling and the Vitamin C Cancer Controversy

We would probably have never heard of the claim that large doses of vitamin C (ascorbic acid) could cure cancer if the claim had not been associated with the name of Linus Pauling. One of the world's most famous scientists, Pauling, who died in 1994, won the Nobel Prize in Chemistry in 1954 for his fundamental work on the nature of the chemical bond and the 1962 Nobel Peace

Prize for his antiwar crusades. Pauling has many discoveries to his name, but was famously beaten by Watson and Crick in the race to find the structure of DNA. Pauling's championing of vitamin C in 1970 is often taken to signify the start of his decline as a scientist. But this is to do Pauling a disservice. His work in the area continued to follow the reductionist tenets and experimental methods of molecular biology, and ironically his vitamin C cure smacks of the medical "magic bullet" so decried by holistic medicine.

Pauling's advocacy of vitamin C can be traced to his own earlier acclaimed discovery that the genetic disease called sickle-cell anemia is caused by an inherited molecular defect. He reasoned that all humans suffer from another genetic disease, hypoascorbemia, or vitamin C deficiency of the blood. He suggested that at some stage in their evolution humans must have lost the ability to synthesize this essential nutrient through a mutation in their DNA. Pauling generalized this radical approach toward disease and its causes in 1968 by proposing a new branch of what he called "orthomolecular" (literally "right molecule") medicine that he defined as "the achievement and preservation of good health and the prevention and treatment of disease by regulating the concentration of molecules that is normally present in the human body. Important orthomolecular substances are the vitamins, especially vitamin C" (quoted in Richards, 37).

There is no doubt that this intrusion by a nonphysician into the realm of medicine caused great furor among the medical profession. Pauling's ideas came under attack from psychiatrists, physicians, and nutritionists as unscientific and unfounded. Pauling was forced onto the defensive and was unable to interest even his colleagues at the Stanford Medical School in his research plans for this new area of medicine. In 1973 he resigned from Stanford to set up his own independent research institute for orthomolecular research. The sociology of fringe sciences shows how scientists can be pushed further and further toward marginal institutions as they lose their resource base within orthodox science.[2] Pauling soon found himself unable to obtain major research grants from the NSF and NIH—the standard sources—and was forced to turn for funding to the very holists and alternative medicine groups he was loathe to associate with. Unsurprisingly, this was taken to confirm the critics' suspicions that Pauling indeed was a closet holist and alternative medicine enthusiast.

Pauling first advocated vitamin C as a cure for the common cold. In his 1970 book, *Vitamin C and the Common Cold,* he suggested in passing that vitamin C might also prevent or cure cancer (he was given the idea by vitamin C enthusiast and industrial chemist Irwin Stone). The famous "War on Cancer" launched by President Nixon was just getting started, and cancer had become *the* public health problem in the United States. It was the second leading cause of death (only heart disease killed more Americans), and accounted for approximately one in five deaths with over a million new cases of cancer treated each year.

Dr. Ewan Cameron

The research on vitamin C and cancer was actually carried out by Pauling's collaborator, a Scottish surgeon, Dr. Ewan Cameron. In his well-received 1966 book, *Hyaluronidase and Cancer,* Cameron set out a new approach to cancer based upon controlling its invasiveness rather than destroying cancer cells. Cameron suggested that an enzyme known as hyaluronidase was released by cancer cells and that this modified the "ground substance," the jelly-like material in which all cells are embedded. The malignant cells could proliferate and penetrate the ground substance and invade surrounding tissues. He proposed that in healthy cells the invasion was kept in check by the existence of a physiological hyaluronidase inhibitor (PHI). Malignant cells and their descendants continuously produced hyaluronidase, and the excess of this enzyme overwhelmed the counterbalancing PHI. Cameron could

91

explain many features of the development of cancer with his PHI idea. Furthermore if PHI could be identified, it could be used in treatment to neutralize the malignant capacity of cancer cells.

Cameron's long-term goal was to persuade a large drug company to isolate and define the PHI substance and prepare it in a form suitable for clinical use. With the added impetus of the war on cancer, U.S. researchers became interested in Cameron's new approach. Cameron himself began treating terminally ill cancer patients with a cocktail of hormones—at the suggestion of a U.S. researcher (not Pauling) he added megadoses of vitamin C. His hope was that the drugs would help render the host ground substance more resistant to malignant invasion and thereby help control cancers. Cameron first wrote to Pauling in 1971 with his initial results. He described them as "very encouraging indeed" (Richards, 77). He felt that the vitamin C had made such a difference "that I have omitted the hormonal regime altogether and am trying to devise a suitable therapeutic program using ascorbic acid alone" (Richards, 77–78). Cameron even wondered whether the structure of PHI might actually include ascorbic acid molecules. If this were the case, given sufficient ascorbic acid, the body would be capable of synthesizing its own PHI.

Cameron was confident enough in his theory and early results to suggest to Pauling that "we could soon cure cancer" (Richards, 79). He knew his approach was controversial because it said, in effect, that the cure for cancer had lain under the noses of specialists for years in corner drugstores and local health food shops. Eventually he planned to publish his findings in the UK medical journal *The Lancet,* and he asked Pauling to help publicize the theory and put it on a "sound scientific basis." Pauling, who at this stage of his career was still at Stanford and trying to interest his colleagues in testing vitamin C on cancer patients, was excited by what Cameron had already achieved and impressed by the new therapeutic successes.

The Vale of Leven Study

Cameron at this point had only seven weeks worth of data on eleven patients. He had administered five grams of vitamin C intravenously for five to seven days, followed by two grams administered orally. Once he saw that patients could tolerate these doses, he had increased them to daily intravenous fusions of ten grams for a week or more followed by daily oral doses of eight grams. What did Cameron expect these massive doses of vitamin C to achieve? He repeatedly stressed that he did not expect to cure cancer, but rather to control it. In other words, the cancer cells would be disarmed and prevented from spreading but would not be killed. Thus, malignant masses would remain even if the treatment was successful, but it was hoped that further growth would be arrested. Preexisting tumors would become benign and isolated, and secondary features of cancer like pain, weight loss, and hemorrhage would be brought under control. His results already indicated that high doses of ascorbic acid could be tolerated with no ill effects and that some of the more distressing symptoms in these terminal patients could be alleviated. In one patient he thought there was some evidence of tumor regression. The results, although inconclusive, showed some promise for the new therapy.

Pauling urged Cameron to set up a larger and more systematic study at the Vale of Leven Hospital, where Cameron worked, and suggested that he employ larger doses (as much as fifty grams per day). As Cameron pondered whether to launch the larger study, he encountered a clinical setback—three of his patients from the initial group who had been doing well suddenly died. He felt that despite the setback there were some encouraging signs that the new treatment had helped ease their disease and that he should continue because the theoretical argument in favor of treatment with ascorbate justified its therapeutic evaluation. With the aid of the princely sum of 4,000 pounds from the Scottish Home and Health Department and 1,000 pounds from the Vale of Leven Hospital, Cameron embarked on a larger pilot study at the beginning of 1973.

As expected, *The Lancet* rejected his paper detailing his initial results. Now he urged Pauling to help publicize his work in the United States. Pauling readily agreed. Using his privileged position as a member of the National Academy of Science, he tried to publish a joint paper with Cameron on the theoretical argument for the treatment of cancer with high-dose vitamin C in the *Proceedings of the National Academy of Science,* where Pauling thought he had guaranteed publication rights. The *PNAS* rather surprisingly rejected the paper on the grounds that they did not publish speculative therapeutic suggestions in such a highly emotive area as cancer (rejecting the paper broke the Academy's own publication rules). Pauling protested, and in the face of the ensuing publicity the editor of the international journal of cancer research *Oncology* (himself an advocate of vitamin C treatment) agreed to publish the paper sight unseen.[3]

Meanwhile Cameron and his colleagues continued with their pilot study, settling finally on a dose regimen of ten grams per day intravenously for up to ten days followed by ten grams per day orally indefinitely. Cameron was convinced that this was "a valuable remedy in the great majority of patients with terminal cancer" (Richards, 92). He was achieving tumor retardation and prolongation of life in many patients and, perhaps most important of all, he felt he was improving their quality of life.

Cameron described for Pauling the "standard response" in his terminally ill patients. They entered into treatment dying from the relentless progress of their tumor, usually heavily sedated and steadily losing weight. On receiving vitamin C they initially showed no improvement and in fact sometimes got worse, which for Cameron was evidence ruling out the placebo effect. About a week after commencement of therapy, patients began to experience a feeling of well-being, recovered their appetites, and began to gain weight. Bone pain from skeletal metastases eased often to the extent that patients could begin to come off their heavy sedation regimes. Other complications of advanced cancer, such as malignant effusions, respiratory diseases, and jaundice were alleviated or arrested. He reported that standard biochemical indices of malignant activity, instead of rising relentlessly, remained stationary and in many patients gradually fell. This "standstill" phase was of variable duration: in some patients it was temporary, while in others it continued for weeks or months before the patient suddenly succumbed to death. The manner of death was also unusual in Cameron's experience. Instead of the characteristic long, drawn-out decline, the patients went through a "whirlwind" reactivation of their cancer and usually died within a few days.

These findings were so encouraging that Cameron started to compile a paper detailing fifty patient histories. Throughout he was concerned with the ethical aspect of the research. Vitamin C was only to be administered to patients who were deemed untreatable in the opinion of at least one other of the attending physicians. Cameron became ever more convinced that vitamin C brought benefits to patients in terms of quality of life, and this, if nothing else, was "good doctoring." His views were reinforced after visiting the United States for the first time in 1973 (where he also first met Pauling). There he encountered the relentlessly aggressive surgery, irradiation, and chemotherapy regimes that were the standard treatment at the time. He wrote to Pauling, "I do not know what kind of 'results' they are achieving but they are certainly causing much mutilation and human suffering along the way" (Richards, 95).

Armed with a draft of Cameron's paper and a set of x-rays showing tumor regression and bone growth in one fifty-five-year-old man who had received no more treatment than ascorbic acid for

95

six months, Pauling tried to persuade his colleagues at Stanford to initiate trials of vitamin C. After failing to convince them, he turned to the director of the National Cancer Institute (NCI) in the hopes of arranging a double-blind control trial. Again he "failed nearly completely" as the NCI officials responded that there would have to be convincing evidence in animals before any trials were begun with humans. Pauling and Cameron felt the animal trials were unnecessary because they had already shown humans could absorb large doses of vitamin C without ill effect, and in any case results in animals would be misleading, since most animals synthesize their own ascorbic acid. Eventually Cameron and Pauling conceded that if they were going to embark on a big trial treating cancer patients under more favorable circumstances when their cancers were less advanced, they would need to undertake animal trials first.

Cameron and Pauling slowly started to interest the wider biomedical community, if not the cancer specialists, in their work. In view of the prevailing hostility, they now played down the theoretical claims for PHI and instead stressed vitamin C as a supportive cancer theory—one that could help with containment. They published several papers (but all in scientific rather than medical journals) and in March 1974 were invited to present their findings to the famous Sloan-Kettering cancer institute that at the time was undertaking trials of the controversial cancer drug laetrile (made from apricot pits). Sloan-Kettering was under increasing public pressure to avail itself of a wider range of cancer therapies, and so agreed to undertake some initial observations of vitamin C.

Cameron had one spectacular case in his support: that of a forty-five-year-old truck driver who had been diagnosed at another hospital with a lymph cancer known as reticulum cell sarcoma. Because of an administrative delay this patient had not been given radiotherapy and chemotherapy and had instead been fed intravenous vitamin C almost as a stopgap measure. Much to the amazement of the treating physicians, within two weeks the patient became clinically well and able to return to work. He then stopped taking his vitamin C, as he seemed to be cured. What made the case so convincing was that later the man relapsed and again began treatment with vitamin C alone and again responded well and was cured (and was then continuously taking large daily doses of vitamin C). Of course, cancer cases occasionally undergo spontaneous remissions, but in this case the remissions had corresponded exactly with the vitamin C treatment, and Cameron knew of no other cases in the literature where a patient with this kind of cancer had undergone two spontaneous remissions. He considered this so remarkable that he published a paper on this one case alone.

The Sloan-Kettering Study

In January 1975 the first report of Sloan-Kettering's observations on sixteen patients treated with vitamin C became available. The case histories showed no sign of any benefit. Replication would be a continuing problem. Cameron and Pauling believed Sloan-Kettering had been unable to replicate their results because they had not selected the right group of patients and had not started treatment with ascorbic acid early enough. The cancers in the Sloan-Kettering patients were all extremely advanced and had already been extensively treated with radiation and chemotherapy. Readers familiar with the earlier books in the *Golem* series will realize that this is a classic instance of a contested experimental outcome, or what we call the "experimenters' regress." How do we settle whether vitamin C is a cure for cancer? The answer is we do an experiment. But whether vitamin C works as a cure for cancer depends on the skills of the physicians in administering and evaluating the treatment—in other words, it depends on their skills at "experimenting" with this new cancer therapy. If vitamin C does indeed cure cancer, then Cameron has the requisite skills and

97

Sloan-Kettering does not. If vitamin C does not cure cancer, then Sloan-Kettering are the skilled practitioners. How do we find out who has the requisite skills? The answer is we do an experiment to see whether vitamin C cures cancer, and so on. With no independent measure of skill available, the results are indecisive and we are caught in a regress.

Historical Control Matching

To try to bolster the credibility of the Vale of Leven results Pauling had been urging Cameron to carry out a double-blind controlled study. Although common in the United States, these studies were seen as more dubious in Britain because it was considered unethical to deny treatment to a group of patients in the control limb of the study (the same issue that later in the United States became a core concern for AIDS activists—see chapter 7). Instead Cameron set about searching the Vale of Leven Hospital records for a matched group of control patients—those who had tumors of similar types and similar histories but who had undergone conventional treatment with no use of ascorbic acid. This technique is known as "historical control matching" rather than the prospective control matching used in a full clinical trial. After much technical debate about the adequacy of the randomization procedures used (with a matched control it is still desirable to randomly select the patients to be included in the study, and in this case, since some had been assigned by Cameron, the study was not properly blinded), Cameron was able to publish the paper, again with Pauling's help, in the October 1976 issue of the *PNAS*. The paper showed a four-fold enhancement in survival rates over the matched control group.

The paper generated much publicity, and *New Scientist* got wind of its controversial findings. A few days later a BBC broadcast was aired. Soon the British national press picked up the story, followed by favorable stories in the *New York Times* and *Washing-*

ton Post. Cameron, alarmed by all the publicity, soon found himself overwhelmed by letters from families of desperately ill cancer patients. There was also a personal edge to the story. Pauling's wife, Ava, had just been diagnosed with stomach cancer and had undergone surgery. She had decided to refuse backup radiotherapy or chemotherapy and was reportedly on the high-dose vitamin C regime of ten grams a day.

Cameron and Pauling now felt they had turned a corner. A leading cancer journal, *Cancer Research,* invited them to submit a review article; slowly other cancer researchers were starting to pay attention. Pauling and Cameron applied to the NCI for a large grant for more randomized control studies, including animal studies. Pauling lobbied assiduously and was helped by a public statement from Dr. Theodore Cooper, assistant secretary for the Department of Health, Education, and Welfare, who said he thought megadoses of vitamin C were valuable in controlling both the common cold and malignant diseases, and that he himself took large doses. But the news was not all good. Sloan-Kettering decided to abandon its trial of vitamin C after treating twenty-three patients with no obvious response. Pauling again pointed to differences between the Sloan-Kettering treatment and that of Vale of Leven. Cameron and Pauling also offered a new interpretation of the negative results. They felt that the patients might have suffered from something they called the "rebound effect." This was a sudden reactivation of tumor growth produced by curtailing large doses of vitamin C (as had occurred with the truck driver case). Sloan-Kettering in any case was receiving adverse publicity over its studies of laetrile, and decided to withdraw altogether from controversial cancer treatments. The debate with Sloan-Kettering was, however, a taste of the much sharper debate that was to ensue when the Mayo Clinic entered the fray.

Pauling's application to the NCI was eventually turned down. The reviewers argued that the Vale of Leven study had not been properly randomized (the same as earlier criticisms of the study). But because of Pauling's continued protests and public criticism, the director of the Division of Cancer Treatment of the NCI, Dr. Vincent DeVita, agreed to approach one of the premier cancer investigators, Dr. Charles Moertel of the Mayo Clinic, to undertake a proper double-blind controlled clinical trail. Moertel, who was a specialist in such trials, agreed.

Located outside Rochester, Minnesota, the privately endowed and enormously prestigious Mayo Clinic, founded by the Mayo brother surgeons in the 1880s, is synonymous with scientific medicine at its very best. Moertel was a leading cancer researcher, and it was he who had led the NCI-funded team that had officially closed the book on laetrile. His verdict on vitamin C was likely to be similarly definitive.

The First Mayo Trial

Pauling wrote to Moertel in advance of the trial and stressed the importance of using patients whose immune systems were not compromised by earlier courses of radiotherapy and/or chemotherapy. He also stressed the need to continue with the vitamin C regime for a long period—based on the famous truck driver case where early termination of the vitamin C treatment had meant that the cancer returned. In reply Moertel agreed that every effort should be made to duplicate the conditions that existed in Cameron's clinical trail. Cameron and Pauling, however, soon became concerned that the difference between treating cancer patients in the United States and Scotland could prejudice the Mayo trial. At the Vale of Leven, patients who needed chemotherapy were referred to another hospital and excluded from the study. Early indications of the protocol for the Mayo study made no mention of chemotherapy in patient selection. Moertel responded to these worries by pointing out that finding patients who had not been exposed to prior chemotherapy was almost impossible in the United

States. More tellingly, he did not think the problem was crucial because if vitamin C operated by stimulating the immune system, then those patients whose immune systems had been suppressed by chemotherapy might actually receive greatest benefit. Pauling, however, was unconvinced and started to deny that the Mayo study was a proper replication because the patients had already undergone orthodox treatment.

In August 1978 the results of the Mayo study became available. They were negative. Sixty patients who had received ten grams orally of vitamin C a day were compared with sixty-three patients who had received a placebo. There was no statistically significant difference in the two outcomes. The overwhelming majority of these patients, however, had previously received chemotherapy and/or radiotherapy. At first Pauling was not too perturbed by the results—after all he had more or less predicted such an outcome. The subsequent framing of the results, however, as a rebuttal of Pauling and Cameron's claims led to controversy. Pauling thought that Moertel had acted improperly in not putting greater stress on the fact that, unlike in the Cameron study, all the Mayo patients had already undergone some form of chemotherapy and/or radiotherapy. Furthermore Moertel's paper, which was to be published in the *New England Journal of Medicine* under the title "Failure of High-Dose Vitamin C (Ascorbic Acid) Therapy to Benefit Patients with Advanced Cancer," mistakenly asserted that 50 percent of the Vale of Leven patients had in fact received chemotherapy. The true figure was actually only 4 percent. Pauling and Cameron, having read a preprint, immediately contacted Moertel demanding a correction of the 50 percent figure. Moertel agreed, but when he in turn contacted the *New England Journal of Medicine,* he found that the paper was too far advanced in the printing process to be changed. He assured Pauling that every effort would be made to publish a correction at the earliest possible time. But the damage had been done. The press was given to believe that the Mayo Clinic

101

had decisively refuted Pauling. Even getting the correction published turned out to be difficult. The journal had a policy that such a correction could be made only by Pauling in a letter to the journal. Pauling, now in combative mood, refused categorically, arguing that since Moertel had made the mistake, the onus was on him to correct it. The dispute over how the error was to be corrected was eventually settled with an agreed-upon formula whereby Moertel would publish a letter saying he had received a letter from Pauling correcting the error. Much to Pauling's fury, however, this gave Moertel the chance to further frame the error as scientifically trivial. For Pauling the error was crucial, as it showed that the study was not a replication of the earlier Cameron study. Cameron himself was very disturbed by the Mayo article and the accompanying negative publicity—for him it meant that patients who read about it might stop taking their vitamin C.

The relationship between Moertel and Pauling, which up until then had been courteous and respectful, now broke down completely. Pauling threatened to launch a libel suit against a local Rochester newspaper which ran a story under the headline "Mayo Study: Pauling Wrong on Vitamin C for Cancer." To avoid the libel suit the newspaper agreed to publish a letter from Pauling where he mentioned that Moertel had sought advice from Pauling as to how to run the Mayo study but had then ignored the advice not to use chemotherapy treated patients. As angry letters were exchanged in the press and scientific journals, both scientists ended up accusing each other of ethical improprieties. What Pauling regarded as Moertel's continued misrepresentation of the Vale of Leven studies was for him akin to a professional breach of ethics, while for Moertel, Pauling was advocating an unproven cancer treatment. The fight reached a low point when Moertel referred in the *Oncology Times* to the "non-randomized study conducted in the little hospital, Vale of Leven in Scotland." Pauling in response pointed out that the Vale of Leven Hospital is a "large hospital with

440 beds" and accepts 500 new cancer patients each year and that Stanford University Hospital, considered to be one of the leading hospitals in the United States, has only 420 beds! Pauling warned Moertel that he considered his description of the Vale of Leven Hospital as having been used "for the purpose of derogation."

The Second Mayo Clinical Trial

Even before the Mayo results were announced, Pauling was pressing the NCI for a second study which would exactly replicate the conditions obtaining at the Vale of Leven. As a famous scientist, Pauling had powerful allies in Congress and the support of no less a person than President Carter. At this stage the NCI was under some pressure from a Senate subcommittee on nutrition to initiate research on the links between diet and cancer and was therefore vulnerable to political criticism. In March 1980 it was announced that Moertel and the Mayo Clinic would undertake another NCI-funded trial of vitamin C. Cameron was less than enthusiastic about this second trial, having become convinced that Moertel was not a "reliable independent investigator" but was a protector of the "established cancer industry." Cameron also foresaw a serious methodological problem with the Mayo studies— patient compliance. He argued that dying patients in the control group would take ascorbic acid without supervision and thus muddy the comparison between the two limbs of the trial. Moertel had tested for compliance in the first study by randomly monitoring ascorbic acid levels in the patients' urine, but Cameron thought that blood tests were a more accurate measure. Cameron offered to help set up the protocols for the second Mayo study, but he was studiously ignored by the Mayo team.

Pauling's wife, Ava, who had continued to take the vitamin C megadose and remained in reasonably good health, eventually died of her cancer at age seventy seven on December 5, 1981. This was five years after being diagnosed. According to NCI statistics, her five-year survival chance was 13 percent. Pauling was devastated by her death, but what he regarded as her successful response to the cancer using only vitamin C hardened his determination to fight on.

The results of the second Mayo trial were published in the *New England Journal of Medicine* in January 1985. They were again negative and in effect killed off this particular alternative means of treating cancer. The Mayo team had elected to study patients with advanced colon cancer because this was the most frequent tumor type in the Cameron Vale of Leven studies. Since there was no known effective chemotherapy for this cancer, the Mayo team felt ethically justified in not using chemotherapy first. Patients' urine was randomly checked to ensure compliance with the trial protocols. The results were that "vitamin C performed no better than a dummy medication. No patient had measurable tumor shrinkage, the malignant disease in patients taking vitamin C progressed just as rapidly as in those taking placebo, and patients lived just as long on sugar pills as those on high-dose vitamin C. Surprisingly, and perhaps by chance, there were more long-term survivors receiving placebo than vitamin C" (Richards, 144).

Accompanying the results of the study was a guest editorial written by Dr. Robert Wittes, associate director of the Cancer Evaluation Program of the NCI (such guest editorials were standard practice in the case of contentious or important findings). Wittes hailed the trial as definitive and claimed, "It is difficult to find fault with the design or execution of this study. Ascorbic acid was given in the same daily doses and by the same route advocated by Cameron and Pauling" (Richards, 142). Wittes rubbed salt in the wounds by adding that the apparent earlier positive findings from the Vale of Leven had probably resulted from case selection bias.

Although Pauling had requested of Moertel the standard courtesy of seeing the paper before it was published, he was able to see the paper only after Moertel had done the round of all the major

TV studios denouncing vitamin C as worthless as a cancer treatment and the Pauling-Cameron studies as "biased." Scrambling for a response, Pauling was put on the defensive by Moertel's media coup. Perhaps sensing that the stakes had now been raised, he issued public statements charging the Mayo Clinic with making "false and misleading claims" and went on to accuse the *New England Journal of Medicine* and the NCI of condoning a "fraudulent" study. Pauling was said to be considering lawsuits against all three bodies.

The nub of Cameron and Pauling's criticisms of the second Mayo study was to reiterate points they had made earlier in criticizing the Sloan-Kettering pilot trial and the first Mayo trial. They argued that the controls were inadequate (this was Cameron's main criticism). Of the one hundred patients in the trial only eleven had had their urine randomly checked. Of these only six were taking the placebo and one of these according to the Mayo report was found to be excreting over 550 mg of vitamin C per twenty-four hours while the remaining five were found to be excreting "negligible" amounts of 550 mg or less. Cameron immediately latched onto this figure of 550 mg; since cancer patients normally excrete 0–10 mg (normal healthy patients excrete 30 mg). Thus at least two out of six controls excreted vitamin C at two orders of magnitude above normal. For Cameron and Pauling this meant that the controls were clearly ingesting possibly as much as one or more grams of vitamin C per day, and this invalidated the study.

Pauling's own immediate criticism of the study was that the patients had not been given vitamin C for an "indefinite time." Rather, what had happened was that vitamin C was halted once tumor progressions were noted and the patients were then given a highly toxic chemotherapy treatment. Vitamin C was administered only for a median time of two and a half months, whereas in the Vale of Leven studies vitamin C had been administered from

105

the onset of the experiment until death (or for the few long-term survivors until the present). The lifespan data given by Mayo were thus suspect. All they had measured was the effect of vitamin C on tumor growth for the initial interval and hence only this contribution to the lifespan of the patients.

Pauling and Cameron's third and final critique of the new Mayo data was perhaps their weakest, as it depended upon the "rebound effect"—something which the Mayo researchers did not think existed. This was the effect first pointed to by Cameron back in 1973; if large doses of vitamin C were suddenly withdrawn, the patient's circulating ascorbate would fall to well below normal levels and might induce acceleration in tumor growth. It was the same argument they had made earlier against the Sloan-Kettering study. Cameron and Pauling believed that it was quite possible that the sudden withdrawal of vitamin C followed by highly toxic chemotherapy may even have shortened the life of the patients in the second Mayo study.

As the minutiae of the second Mayo study were put under Pauling's microscope, he found more differences. For one thing, Cameron had carried out his work in a hospital and hence had been able to make detailed observations of the initial positive response to vitamin C treatment. The Mayo patients were ambulatory and were not examined during the first month of vitamin C treatment, so the initial improvements documented so assiduously by Cameron went unnoticed.

The underlying gist of Pauling and Cameron's criticism was that the Mayo researchers had failed to grasp that their own approach was about cancer control and that they never claimed, except for a very few fortunate patients, that they could *stop* tumor progression. Rather the claim was that they could slow tumor progressions, improve quality of life, and slightly but significantly expand survival times. The Mayo oncologists had tested vitamin C as if it were a cytotoxic drug to be administered for a short period of time, and whose therapeutic impact is measured primarily in terms of tumor shrinkage.

Pauling was particularly goaded by the second Mayo trial and what he considered to be the underhand way that Moertel had denied him access to the paper until the day of publication. But as far as most reputable cancer researchers, the press, and the wider public were concerned, the game was over. Pauling tried in vain to pressure the NCI and the National Cancer Advisory Board and the editor of the *New England Journal of Medicine* for retractions, but made little progress. His bullying tactics and threats of lawsuits seemed counterproductive. He could see that his funding, which had always been precarious, was being threatened by the negative publicity. Pauling still believed in his personal ability to win over his fellow scientists. He even mischievously offered to visit the Mayo clinic and give a talk on vitamin C and cancer. The offer was politely turned down by a Mayo administrator on the grounds that it would be impossible to arrange a suitable audience for him.

Moertel, who had publicly been accused of fraud by Pauling, retreated into silence apart from the occasional public pronouncement that the study had been properly conducted and that it would have been ethically repugnant to keep patients on a treatment under which they were regressing. Wittes, who had published the supportive editorial, and Pauling maintained a correspondence. Wittes found it totally implausible that vitamin C could have any effect past the point of demonstrable tumor enlargement, and to support this he went back over Cameron's cases to show that even in the Vale of Leven trials there had been no such regression after tumor enlargement. Nor could Wittes accept Pauling's claims about the rebound effect. The debate was inconclusive and Wittes did not change his mind.

With Pauling and Cameron struggling to get papers published in the mainstream cancer journals and failing to obtain retractions from their critics, it was apparent to most people that the issue was dead. Two negative trials had been conducted by the most renowned cancer institute with the backing of all the leading cancer researchers. The Mayo had also given the appearance of bending over backward to respond to Pauling and Cameron's criticisms of the first trial (criticisms which the Mayo scientists themselves found to be "obscure"). The controversy was closed, with defeat for Pauling and Cameron. As in many such controversies, the originators refused to bow out gracefully. Pauling's death in 1994, however, in effect ended the sustained effort to promote vitamin C as a cancer cure.

Conclusion

What then are we to make of this episode—one of the few cases where a nonstandard approach to cancer has been checked by the best instruments of orthodox medicine? One point that is worth bearing in mind is that, as we document in the chapter on AIDS (chapter 7), cancer clinical trial methodologies have undergone evolution precisely because of the sorts of criticism made by Cameron of the Mayo studies. In the AIDS case it became difficult to carry out clinical trials double-blinded, because desperately ill patients did not want to be in the placebo limb of the trial and were sharing drugs with friends and fellow AIDS sufferers. In the AIDS case exactly the sort of matched historical control that Cameron used became accepted. (Though they are still open to criticism and the RCT is still the gold standard). This does not mean that the criticism of Cameron's randomization techniques might not still be made.

Also important in shaping the debate and in particular the issue of replication were the differing cancer treatment regimes in operation in the United States and Britain. The administration of cytotoxic chemotherapy as applied in the United States was not standard practice in Scotland. And Cameron's own choice of pursuing a historically controlled trial rather than an RCT was affected by the prevailing norms in the UK, where RCTs were less common.

Are the Mayo studies definitive, then? As we have seen, this is a case of experimenters' regress. But the argument has in effect been closed in favor of orthodoxy. Experiments alone did not settle matters, but given the implausibility of Pauling and Cameron's claims within the orthodox framework of cancer theory and practice, the experimental evidence offered a credible source of rebuttal. Cameron's own experimental evidence was said to be methodologically flawed. The publishing difficulties which Cameron and Pauling experienced meant that at no stage were they able to get their clinical results published in an orthodox medical journal. On the other hand, their Mayo clinic opponents twice published their results in the prestigious and powerful *New England Journal of Medicine*.

On offer are two versions of the impact of vitamin C on cancer and how it should be assessed. The dominant professional view of the leading cancer researchers is to assess vitamin C according to their standard methodologies. Within their framework and view of medical expertise, vitamin C is not effective. Cameron and Pauling's claim, however, that vitamin C is a control rather than a cure for cancer, does not seem to have been refuted. Nevertheless, given that interest in Pauling and Cameron's work was in effect killed off by the Mayo studies, neither is there a mass of evidence supporting the use of vitamin C as a cancer control; their claims continue to live on in the twilight world of alternative medicine. Pauling and Cameron failed to get the medical establishment to change their way of evaluating cancer drugs, and their own particular claims for vitamin C remain unproven.

The messy fight between Pauling and his critics seems to be typical of these sorts of cases. It was never conclusively shown, in a quasi-logical sense, that the vitamin C treatment would not alleviate the symptoms of cancer and would not enrich, and perhaps extend, the lives of sufferers. Assuming that the results of the experiments were meaningful, what was shown was that those who had already been treated with chemo- or radiotherapy would not benefit, nor would those who were given vitamin C for a relatively short time. So if vitamin C worked at all, and this was never decisively demonstrated either, it worked only on a subset of cases.

It is tempting to reach too quick a conclusion from this typical case of divided expertise, namely, "more research is needed." Perhaps in this case it would be the right conclusion, perhaps not. The trouble is that what we already know about science (e.g., see *The Golem* and *The Golem at Large*) shows us that *any* piece of scientific research which is examined closely enough will reveal the same deficiencies when compared with quasi-logical standards. Therefore, the liberal-minded approach is a prescription for more research on everything, and that is no kind of advice at all in a world of scarcity. Every choice to do more research has an opportunity cost for some other project.[4]

Evelleen Richards, the author of the study from which we have taken most of our description of the vitamin C debate, recommends a change in the way alternative medicine is assessed to something like the approach followed in the Netherlands. The Dutch have instituted a system which gives a far greater weight to consumer preference in the distribution of medical resources. Given the huge consumer demand for alternative medicine, this means more money for such cures. In turn, this means that treatments are supported by the state even when there is no scientific evidence that they work or even an unwillingness to expose them to standard scientific tests. Richards too believes consumers have a right to express their preferences for unproven remedies and influence the distribution of state resources accordingly. We believe this conclusion confounds what is reasonable for medicine as succor with what is reasonable for medicine as science.[5] Medicine as science and as a collective responsibility must not be driven by popular opinion, even though individuals may be right to seek out any unproven alternative for themselves when the ultimate questions are asked.[6] To say "leave it to the people" is to risk the abrogation of our long-term collective responsibility to scientific medicine, even if sick and dying individuals might still be wise to try the cure.

Yuppie Flu, Fibromyalgia, and Other Contested Diseases

It is 1934 and polio is still ravaging the United States. In California, following three years of decline, there is a severe epidemic, with 1,700 cases reported in the Los Angeles area alone. School assemblies and fairs are prohibited, beer parlors are encouraged to enforce hygienic procedures, and housewives are warned that "dust is a germ carrier" and that they should "use vacuum cleaners for sweeping rather than old-fashioned brooms." Fear is in the air.

At the Los Angeles County Hospital (LAC), where most of the suspected polio cases have been taken, physicians stand guard and question all incoming patients. The staff of the contagion ward are continuously monitored for symptoms of the disease. In May health care workers at LAC begin to get sick. By December 198 staff (4.4 percent) are reported as suffering from polio. In an effort to stem the epidemic, convalescent serum is administered to the entire hospital staff.

But the characteristics of this polio outbreak are different from previous ones. An unprecedentedly high number of cases are adults. It is also unusual for polio to break out at a hospital—there has been only one known mass outbreak at such an institution before.

What was happening at LAC and whether it was really a polio epidemic soon became the subject of a detailed investigation. Adding pressure to the inquiry were health care workers who sought compensation for their illnesses. The findings of the U.S. Public Health Service investigation were perplexing. In twenty-five cases, chosen at random, no definite paralysis or spinal fluid abnormalities were found. It even proved impossible to calculate the traditional polio statistic—the ratio of paralytic to nonparalytic cases. Medical charts revealed only minor motor impairments detected after a vigorous neurological screening. Yet the patients themselves felt ill and sought the usual orthopedic treatment. As noted polio researcher and historian John Paul observed, during the epidemic the poliomyelitis ward at LAC "had the appearance of a ward occupied by patients who suffered extensive trauma inflicted in a disaster area, whereas in actuality very few patients turned out to have any paralysis at all" (Aronowitz, 19).

What then did these apparently ill people suffer from? Some have suggested that they experienced a collective form of hysteria. As one investigator wrote at the time, "Probably I see and am in contact with 100–200 polio patients a day—but remember hardly any of them are sick. . . . There is a hysteria of the populace due to fear of getting the disease, hysteria on the part of the profession in not daring to say a disease isn't polio and refusing the absolutely useless protective serum" (quoted in Aronowitz, 23). Although the majority of the LAC cases fully recovered, a subset, including a group of nurses, suffered from prolonged and recurrent symptoms. Their complaints and their attempts to win permanent disability payments kept the epidemic in the public consciousness for years to come.

In the 1950s researchers revisited the LAC case and concluded that it shared features with other epidemics in other places and countries that had nothing to do with polio. They identified the disease as a new syndrome where no particular set of symptoms is uniquely caused by a specific infectious agent. The syndrome was renamed "benign myalgic encephalitis"—*benign* because no one dies, *myalgic* because of the diffuse muscle pains experienced, and *encephalitis* because subjective symptoms were thought to be secondary to brain infection and inflammation. This new syndrome was the first of a class of such syndromes which have remained controversial throughout their history.

Chronic fatigue syndrome (CFS), disparagingly referred to in the 1980s as "yuppie flu," is the best known of these sorts of diseases. It was called yuppie flu because of the preponderance of wealthy young Californians in the initial pool of afflicted people and the continued doubts as to whether it was a real disease. Other examples are "sick building disease," "Gulf War syndrome," "repetitive strain injury," and "irritable bowel syndrome." The latest to join this list is "fibromyalgia"—persistent muscle pain throughout the body. It is said to afflict over six million Americans (90 percent of whom are women), four times the number that will develop cancer in a year.

Such diseases first appear as a collection of symptoms shared by groups of people. The symptoms are hard to explain in terms of any known physical disease cause. Such diseases, because of their mysterious nature, often attract media attention. Patient advocacy groups often form around the disease, lobbying for more medical research and to get the symptoms recognized as a real disease. Indeed, it was the afflicted nurses at LAC campaigning for disability rights that helped bring about the new diagnosis of benign myalgic encephalitis. Even when such diseases are recognized by the medical profession, persistent doubts remain as to their reality. Symptoms are attributed to psychosomatic causes—"It's all in the head." The vague lists of symptoms, lack of definitive tests, and nonexistence of a physical cause make such diseases hard to identify. They are unlike, say, SARS, strep throat, or a broken leg, where on the face of it either you have the illness or ailment, or you don't have it, and there is a widely accepted method of diagnosis based on the idea that a specific virus, bacteria, or pathology is the cause. The diagnosis may be fallible, but the existence of the disease is not in question.

As mentioned above, patients play a prominent role in trying to get these less well-defined diseases accepted. Patients become not only advocates but on occasion claim an expertise for themselves greater than that possessed by medical professionals. Patient groups make strong claims. One patient advocacy group in the case of repetitive strain injury proclaimed, "We're the experts: not the doctors, or the consultants, or the physios. It's us. We're the ones who have to live with [RSI] day in, day out. It's us they ought to be asking if they want to find out about RSI" (quoted in Arksey, 2). But how much expertise can lay people acquire in defining and understanding diseases? It is this question which underlies our discussion in this chapter.

Chronic Fatigue Syndrome

The first symptoms of what became known as "chronic fatigue syndrome" (CFS) were reported in the early 1980s. Physicians described patients with a lingering viral-like illness that manifested itself as tiredness and other largely subjective symptoms. Initially it was thought that the cause was Epstein-Barr virus (EBV), a form of herpes virus that persists in the body after acute infection and which might cause recurrent symptoms. Isolated cases of recurrent EBV infection had been reported over the previous forty years. But proving that EBV is the cause of chronic fatigue syndrome is complicated due to the widespread exposure of the public to EBV. Many perfectly healthy people possess EBV antibodies.

In 1985 a cluster of over a hundred cases at Lake Tahoe in California finally attracted the attention of the Centers for Disease Control (CDC). Local doctors found large amounts of EBV antigens in their patients. This outbreak spurred a headline in *Science* magazine, "Mystery Disease at Lake Tahoe." From the very beginning the issue was controversial, with some Tahoe physicians deeply skeptical as to whether an epidemic was occurring at all. As one doctor commented "They think they notice something, then they start seeing it everywhere" (quoted in Aronowitz, 25). CDC investigators, following standard epidemiological practice, created a case definition and intensively monitored fifteen Tahoe patients. Although they did find some serological anomalies, there were overlaps with the control group and serological evidence for a range of other infections. Their conclusion was that the reported symptoms were too vague for proper case definition and that EBV serological tests were not reproducible enough to be reliable indicators of the disease. They pointed out that a sensitive and specific laboratory test was needed before anyone could be certain that there was in fact a disease epidemic in Tahoe.

Patients who suffered from what they took to be chronic Epstein-Barr virus infection were less circumspect than the CDC. They started to lobby for the acceptance of the syndrome and attended an April 1985 consensus conference on chronic Epstein-Barr virus infection organized by the National Institute for Allergy and Infectious Disease. Despite the skepticism of the CDC and other expert medical opinion, the disease was launched at this meeting; the launch was directed at both the medical and lay worlds. Popular journals focused on the new disease, private laboratories promoted EBV blood tests, and patients began to arrive in droves. For a while the disease was treated as real by the medical establishment. An editorial in a leading journal of allergy and immunology declared that the "syndrome of Chronic Epstein-Barr virus exists" (quoted in Aronowitz, 25). Much of the credibility and legitimacy of the new disease rested upon the fact that EBV was a known disease with a well-understood pathobiological mechanism and a diagnostic test—the EBV serological test.

By 1988, however, the reliability of the EBV serological test had come into doubt. Another consensus conference was held. The outcome was to rename the disease as "chronic fatigue syndrome" (CFS) and to offer a way to diagnose the new syndrome. Typically a patient should have had a chronic and disabling fatigue for at least six months' duration with the exclusion of any other explanation. The EBV serological test was now no longer regarded as definitive. Instead diagnosis took the form of a "Chinese menu," with a positive diagnosis needing to meet two major plus any eight of fourteen minor criteria. Symptoms included headache, myalgia, chest pain, and joint pains. The new definition was immediately criticized as arbitrary, but now the disease took on a life of its own, as all doctors and patients had to do was apply the checklist of symptoms.

Doubts remain to this day as to whether CFS is a real disease. Studies have been carried out pertaining to show that the disorder is psychological in origin. But these studies have in turn been criticized on methodological grounds and also suffer the difficulty in imputing the direction of causality. It seems likely that someone suffering from an as yet undiagnosed medical condition for many years would indeed suffer psychological effects. One study attempted to cast doubt on the disease by conducting a randomized, double-blind, placebo-controlled test on the effect of treating sufferers with an antiviral drug, acyclovir, known to be active against the herpes virus. The study found that the drug had no advantage over the placebo. Although the authors of this study framed it as evidence against the EBV hypothesis and the very existence of the disorder, the study has been criticized (including by patient advocacy groups) as too small and methodologically weak because the subjects were not typical of most patients suffering chronic fatigue syndrome. In short, although much research and several

117

international conferences have focused upon CFS, there is no consensus as to whether it exists as a physical disease with a biological cause and a pathobiological mechanism. Doctors seem to recognize increasingly that part of the problem of the disease is that it has a psychosociopathological element—that is to say, because patients think they have the disease, they act as if they have it, and then they do indeed experience the symptoms as real. This is an example of the complex interaction between mind and body which we discussed in chapter 1 on the placebo effect. Indeed, if CFS is framed as psychosociopathological in *origin* it can be thought of as a "reverse placebo effect." Rather than the mind curing the body by thinking that a nonexistent cure is efficacious, the mind harms the body by thinking a nonexistent disease is a real disease.[1]

Fibromyalgia

The arguments concerning CFS are very similar to those which have surrounded the existence of fibromyalgia. This new disease entered the medical lexicon in 1990. Its name is derived from the Greek *algia*, meaning "pain," *myo*, meaning "muscle," and the Latin *fibro*, indicating the connective tissue of tendons and ligaments. It designates a condition of persistent muscle pain throughout the body that is often accompanied by other symptoms such as tiredness and insomnia, diarrhea and abdominal bloating, bladder irritation and headache. Many cases occur after a traumatic event such as surgery, viral infection, physical injury, or emotional trauma, but others have no known cause. Dr. Frederick Wolfe, the director of the Wichita Research Center Foundation in Kansas, was one of the first doctors to help define this new disease. Since the 1970s he had observed more and more patients with diffuse muscle pains but with no evidence of inflammation or muscle pathology. In 1987 Wolfe brought together twenty Canadian and U.S. rheumatologists who had been observing similar symptoms. The new disorder of fibromyalgia was born. A simple diagnostic test, endorsed by the American College of Rheumatology, was developed. The test involved a physician pressing firmly on eighteen designated points where muscles and tendons were attached to bones. A patient who felt pain at eleven or more of these points was declared to have fibromyalgia.

The following account by a *New Yorker* magazine journalist shows how fibromyalgia is perceived from the point of view of the patient. The patient (referred to as Liz), is a friend of the journalist. She is a fifty-one-year-old, recently divorced woman who teaches at an elite New England college. Liz's problems began in 1994 when she underwent surgery for a sinus infection. She failed to recover and suffered tiredness, insomnia, and muscle aches: "My internist told me that it was all tension, that I'm middle-aged, and reacting to the stress of raising two kids, five and eight" (Groopman, 82). No one was quite able to explain Liz's condition: "Liz had had episodes of depression in the past, but this felt very different. One specialist she consulted thought that her pituitary gland might have been nicked during the sinus surgery, but extensive endocrinological testing showed otherwise" (Groopman, 81).

After several failed attempts to diagnose the disease—for a while her symptoms were explained as a rare food allergy—she was given the diagnosis of fibromyalgia and chronic fatigue syndrome. Liz's experience of being shunted from doctor to doctor is typical. In the era of managed care, doctors have neither the time nor the incentive to listen to a seemingly endless list of inexplicable symptoms. Fibromyalgia patients often set off a game of clinical hot potato, with each doctor eager to pass the patient on to a colleague as quickly as possible. One doctor referred to these patients as the "bane of the medical profession" (Groopman, 81).

There is as yet no cure for fibromyalgia. In desperation Liz turned to alternative medicine: a Vietnamese monk performed acupuncture to no avail; a chiropractor diagnosed a damaged neck

119

from a teenage car accident; an osteopath prescribed painkillers for the rest of her life. Liz, now even more desperate, went back to an internist:

> "The first thing I said to him was 'You need to believe I am really sick, not just complaining.'" For her fatigue she was given Ritalin, and for her insomnia she was given Ambien. . . . Recently she has been taking Prozac but it has not made much difference. She closely follows reports on the Internet and in fibromyalgia and chronic-fatigue newsletters, searching for possible solutions. "I've tried everything" she went on. . . . Finally last year, she gave in and took time off teaching because of the pain, fatigue, and episodes of what is commonly referred to as "fibrofog," or an inability to think clearly. (Groopman, 86)

Liz talked about the nomenclature of the disease, and about how fibromyalgia seemed to be merging with chronic fatigue syndrome: "'Chronic fatigue has become a humiliating term—the yuppie disease, begging to be laughed at,' Liz said. 'Fibromyalgia is more socially acceptable'" (quoted in Groopman, 86).

As with chronic fatigue syndrome there is a strong medical lobby who dispute the existence of the disease. They make the now familiar argument that classifying the symptoms as a new disease may do more harm than good. Frederick Wolfe, who first identified the disease, now shares this view. "For a moment in time, we thought we had discovered a new physical disease. . . . But it was the emperor's new clothes. When we started out, in the eighties, we saw patients going from doctor to doctor with pain. We believed that by telling them they had fibromyalgia we reduced stress and we reduced medical utilization. This idea, a great, humane idea that we can interpret their distress as fibromyalgia and help them—it didn't turn out that way. My view now is that we are creating an illness rather than curing one" (quoted in Groopman, 89). Wolfe's experience was that the number of tender points he found in fibromyalgia patients correlated with their degree of unhappiness!

Symptoms that are routine become intensified because of the very fact that there is now a disease within which to fit them. Critics point out that a third of healthy people will have aches and pains in their muscles at any one time, and a fifth will report significant fatigue. Furthermore nearly 90 percent of a general healthy population report at least one somatic symptom, like headache, joint ache, muscle stiffness, or diarrhea, in any two- to four-week period. A typical adult will thus have one symptom every four to six days. For people prone to fibromyalgia the everyday somatic symptoms become a growing focus of attention. As Dr. Arthur Barsky, a Harvard professor of psychiatry, puts it, "They become trapped in the belief that their symptoms are due to disease; with future expectations of debility and doom. This enhances their vigilance about their body, and thus the intensity of their symptoms" (quoted in Groopman, 86). Barsky also notes the role of powerful groups with vested interest in the disease, "including doctors and other practitioners who run clinics, lawyers involved in disability litigation, and drug companies marketing treatments of unsubstantiated benefits" (quoted in Groopman, 87). Fibromyalgia turns out to be a very convenient diagnosis for lawyers to use in arguing for disability, since the disease is heavily dependent on self-reporting. One study of 1,604 patients at six medical centers indicated that more than a quarter of fibromyalgia patients received disability payments.

There is no doubt that fibromyalgia is one of the most contested disease categories in current medicine. Many of the doctors the *New Yorker* journalist spoke with refused to give their views on the record. Some worried that any hint of empathy with the victims would lead to a deluge of referrals; others worried that voicing skepticism about the syndrome could make then vulnerable to public attack. One well-known critic of the disease reported receiving over two hundred pieces of hate mail and being attacked by fibromyalgia advocates on the Internet and in newsletters.

Patient Advocacy

We now turn to examine the role of patient advocacy in defining these new disease entities. Patient advocacy seems to have been born with the successful activities of AIDS activists in the 1980s (chapter 7). In the case of CFS many patient advocacy groups are explicitly modeled on the AIDS activist groups. In the UK the CFS activist groups include ME Action, the ME Association, and the National ME Center. In the United States the Chronic Fatigue Immune Dysfunction Association (which publishes its own newspaper the *CFIDS Chronicle*) has played a prominent role. The very names of these groups reflect in part the battle over the disease. The UK groups' use of ME in their titles reflects their overriding concern to get their disease recognized as a real medical condition (caused by brain inflammation as in encephalitis) rather than just a collection of symptoms. The main U.S. group's use of the word "immune" signals the link to AIDS and all the attention that that disease has received. The naming of the disease as an immune condition points to some underlying disruption of the immune system causing particular individuals to succumb. Many patient activist groups present AIDS and CFS (and other disorders) as only the tip of an iceberg of previously undiscovered immune conditions.

Patient advocacy groups lobby to change health policy about their disease. Often their representatives make the journey to Capitol Hill to testify before congressional committees. In the UK, CFS patient groups have been active in responding to an influential 1996 report on CFS published by the prestigious Royal College of Physicians, Psychiatrists and General Practitioners. The report concluded that CFS was neither purely physical nor purely psychological but arose out of a "complex interaction between the mind, the body and the patient's social world."[2] The UK patient organizations immediately published a critical response accusing the report of bias in favor of "psychiatric models of causation and treatment." They followed up with a campaign to try to discredit its findings organized around the slogan "Fighting for Truth" or "F for T." A petition was delivered to Parliament calling for the report's withdrawal. They submitted background papers which argued against the psychological framework for understanding and treating CFS. In their view psychiatric disorders associated with CFS, such as depression, were caused by a real viral infection which researchers should be seeking to identify.

Patient advocacy groups use the individual testimony of patients to great effect. Indeed, it is hard to question the reality of a disease when confronted with someone obviously in pain who claims that her disease is being ignored by the medical profession. Having a name for the disease and identifying it as something real are often empowering—a way to deal with what appeared to be a hopeless condition. According to Robert Aronowitz, a physician turned sociologist who has studied these advocacy groups, "A successful young woman suddenly develops a mysterious debilitating illness. Because her physicians are unable to make a precise diagnosis, they become impatient and suggest that the problem is psychological. Friends and family become more frustrated with the patient's situation and begin to lose interest and sympathy. When all hope appears to be gone, the patient is diagnosed as suffering from myalgic encephalitis or chronic fatigue syndrome either because she discovers the diagnosis herself or meets a knowledgeable and compassionate doctor. With the disease named and some time elapsed, the patient begins to recover, often delivering a moral lesson in the process" (Aronowitz, 33).

The rhetoric of advocacy combined with patients' (often overlooked) subjective experiences of their illness can evoke a bitter struggle between patients and the biomedical establishment. One lay advocate of CFS warns against "the rigid mind-set of those who have tried to submerge the illness as a clinical entity, discredit the

123

physicians who have stood by us, and demean those of us who have CFS" (quoted in Aronowitz, 34). The medical establishment is often accused of incompetence and even conspiracy. One patient advocate accused the CDC of covering up a national epidemic of immune system dysfunctions and viral disorders. One of the two Tahoe physicians who originally interested the CDC in the Tahoe outbreak was depicted as having been "run out of town" because of the epidemic's possible damage to the local tourist industry.

Economic interests, however, can cut both ways. Compensation for disability has been a central concern for those suffering with CFS. The CFIDS association encourages patients to claim such compensation. Commercial labs also have a strong economic interest in the disease and promote serological tests, which result in more diagnoses, more tests, and more money for them.

Some of the activist groups publish summaries of relevant scientific and popular reports and scan such reports for usefulness to their cause. The groups seem to have the ability, familiar from the AIDS activists, to assess and critique scientific studies. Such critiques may be methodological, or refer to the quality of the argument or the ideological motivation of the authors. One group keeps a roll call of physicians sympathetic to CFS and will on occasion "out" researchers whom they take to be overly critical of CFS. When a prominent researcher, Steven Straus, published two studies viewed as damaging to the disease's legitimacy, the group organized a drive to have the NIH dismiss him (they failed).

The same tensions over the status of science that exist among the AIDS activists (chapter 7) can be observed with CFS. Although these activists often criticize what they take to be the unwarranted power of medical science and postulate a conspiracy among the medical elite, ultimately they want the medical establishment to legitimate their disease using the methods of science and proclaim it to be just an "ordinary" disease. Indeed, the defining feature of patient advocacy is their attempts to become well enough versed in medical and scientific terminology to challenge the medical establishment and on occasion to do the science for themselves. But how far can a lay person go in learning to become a scientist?

Becoming a Scientist

Medical science involves its subjects in a way that few other sciences do. As explained in the introduction and chapter 3, so long as the patient is not unconscious, it is likely that self-reports will form a big part of a diagnosis. Furthermore, it may often be the patient alone who knows if he is better. Thus doctors have to rely on their patients both in deciding what is wrong and deciding if they have succeeded in putting it right. A lot of the time, the patient is a partner in the doctor's medical procedures whether the doctor likes it or not. The reporting role for the patient shades over, then, into something more like participation. In the cases we have discussed more than shading is going on; here patients take the lead in defining and establishing the existence of new diseases. Patients become, or try to become, scientists.

There are other ways for patients to become medical scientists. Anyone suffering from a chronic illness such as diabetes is likely to become her own day-to-day diagnostician and pharmacist. Diabetics become experts at understanding and maintaining their own blood sugar levels. Another way in which an ordinary person may become a real or quasi-medical scientist is in the case of esoteric or illegal uses of drugs. One such group are bodybuilders who use drugs for recreational purposes in building up muscle tissue. A recent study of this group by sociologist Lee Monaghan offers insights into how far lay people can go in acquiring scientific expertise.

Setting aside the more destructive of these activities, we can look at bodybuilders as a group who at least consider that they

125

have more detailed understanding of their own needs, and their own physiologies than the typical doctor. Monaghan describes the world of bodybuilders as a subculture that possesses a remarkably detailed body of folk pharmacological knowledge (ethnopharmacology, as he calls it). For instance, knowledgeable bodybuilders refer to "receptor sites" which, drawing upon the biochemical model of the body, are specific areas within the internal cellular structure of the body which are open to the various chemical messages transmitted by ingested or injected steroid molecules. Here is Bill talking about his steroid regime:

> Anapolon has got 50 milligrams and Anavar has got 2.5 milligrams. It all depends on the receptor sites for these particular steroids. Anapolon has got very limited, small receptor sites to hit which is why it needs 50 milligrams—to hit the receptor sites. Oxandrolone—Anavar—has obviously got very easy receptor sites, the receptor sites are very open so less is required. This is normally the difference. People sort of think: "oh the stronger the better," but it's not. This is why Anapolon shouldn't be taken really because it's quite toxic. Of that sort of 50 milligrams you might hit [the receptor sites with] sort of maybe 10 or 20, but then you've got 30 milligrams wooshing around your body, looking for the way out. You know? (Monaghan, 111)

According to bodybuilding "steroid guru" and self-proclaimed "lab rat" Dan Dochaine, there is no scientific or medical research on the most effective way to use steroids for athletic enhancement. The knowledge seems to be shared among bodybuilders in the gyms and, more recently, on the Internet, in addition to some who read and critically translate relevant medical literature. Here is John, an everyday participant (without any formal medical credentials), talking about the progressive loss of sensitivity of receptors to an exogenous steroid: "So I thought I'd read up on it. Other people talk to you about it, people who know about it, like. And they say that if you do stay on the same thing . . . after about six weeks, it's not going to work for you anyway, no matter what it is. Cause your body gets used to it. Your receptors won't accept it any more. So after it's about six weeks, you're better off changing" (Monaghan, III).

We are not in a position to say whether the bodybuilders' knowledge is sound enough to enable them to mold their physiologies with as little harm as they seem to think. According to Monaghan, bodybuilders present themselves as the most sophisticated of all athletes in their steroid use. He observes that there is "a systematic and qualified basis to their shared ethnoscientific reasoning." But one cannot deny that, whether they are getting it right or wrong, they are likely to know more about the detail of the matter than the average physician.

The bodybuilders are an esoteric group. The knowledge that they believe they have is applied to themselves alone. They have no ambitions to have muscle-enhancing drugs offered at the public expense or to encourage more publicly funded research on pharmacological aids to their sport. In a sense, the bodybuilders, though they share their knowledge as a group, can be thought of as individuals when we set them in our framework of the tension between the individual and the collective. They may not be dying (though the media, in reiterating a malevolent assumption, may sensationally claim steroid-using bodybuilders are literally "dying to be big"), but they have chosen a certain lifestyle which leads them to seek out unorthodox treatments on their own responsibility. For this reason there is, again, no deep dilemma for the science of medicine. The medical profession may have a duty to advise, but as a profession it has no pressing decision to make about whether to take on the bodybuilders' knowledge or reject it; at worst there is no more opportunity cost for other treatments than in the cases of, say, smoking or overeating.

When it comes to thinking about how much scientific expertise lay people can acquire, the difficult cases are the ones we discuss in this chapter where patients form themselves into a medical pressure group to establish the existence of a new disease and force the medical profession to accept their definition of the world. In such cases the patients not only become scientists but demand that their new science be given a public imprimatur. These cases, unlike that of the bodybuilders, have much more direct and significant implications for the collectivity. So how should we think about these sorts of cases?

Let us start by reviewing their salient features. The types of disease we have been examining in this chapter are riddled with uncertainties. Medical experts disagree over whether the diseases exist at all and whether they are to be explained by physical causes, psychological causes, or by a complex interaction between the two. The treatment for the diseases (if they exist) is similarly uncertain: some physicians recommend a combination of therapy, exercise, and lifestyle changes; others recommend drugs; and yet others, having no cure to offer, try to pass on these sorts of patients to other doctors (the hot potato syndrome). It is into this arena of uncertain knowledge which "lay experts" have entered. But have their role and actions always been beneficial? And how expert are the so-called lay experts?

Classifying diffuse and uncredentialed lay groups in terms of their expertise is notoriously difficult. But some sufferers do have credentialed expertise. For instance, in the case of the LAC outbreak many of the afflicted were themselves health workers— nurses and even doctors. It is likely, given the prevalence of the diseases we are discussing in this chapter, that other health workers are among the sufferers. Patient advocates also often draw upon scientific training in other fields. For instance, a training in statistics or a science such as psychology, which heavily relies upon statistical evidence, may help in understanding and interpreting epidemiological data. But given that expertise is highly bounded, it seems unlikely that general types of scientific expertise can make people experts qua the investigation of the etiology of particular diseases. It is rather like expecting a molecular biologist to be an authority on the merits of string theory. Of course, a general level of medical competence and/or scientific training may enable people to bone up on the new area and acquire what we have called interactional expertise, but the barriers to making an actual contribution would seem to rule out all but the most extraordinarily well placed. There will, however, always be exceptions. For instance, Mary Bigler, head of the LAC contagion unit, herself became ill in June 1934. She later went on to coauthor one of the major epidemiological reviews of the outbreak.

The movie *Lorenzo's Oil* (based upon real-life events) also reminds us that given enough motivation, lay people without any medical qualifications can acquire enough expertise to learn medical terminology and effectively read, comment upon, and criticize the medical literature and even make contributions to medical science. The drama of the movie is built around a World Health Organization employee whose son is diagnosed with the incurable neurodegenerative disease adenoleukodystrophy (ALD). He refuses to accept the standard prognosis that nothing can be done and starts to cull the medical literature and encourage research into metabolic pathways affecting his child's condition. He concludes that with the use of special dietary oils the progression in the disease can be halted—an intervention which entails at least some contributory expertise. This 1993 movie itself has become a rallying call for activist groups.[3] Studies of activists also show the extent to which concerned sufferers from these diseases can gain enough medical knowledge to make a difference. Hilary Arksey points out in her study of RSI that lay people grouped around the disease, like the AIDS activists (chapter 7), can occasionally contribute by carrying out small-scale research projects which make a genuine contribution to new knowledge.[4] The limits of this contribution are, however, worth emphasizing. Even in the best documented of these cases—that of the AIDS activists—where some

129

of the activists were extraordinarily well-educated and highly motivated, they did not actually get to run clinical trials or publish articles in the mainstream medical journals. Their main role was to attend conferences and debate with and advise the medical researchers as to how to carry out such trials.

There are also cases documented where lay people, by dint of their own experiences in a particular domain, have expertise that scientists and doctors cannot easily acquire. In the case of the Cumbrian sheep farmers responding to the Chernobyl nuclear fallout, discussed by Brian Wynne, the farmers were experts on the ecology of their own farmland and on the behavior of their sheep.[5] This sort of expertise (although not formalized) is similar to the expertise built up by the bodybuilders mentioned above.

Also, as we have reiterated, patients unquestionably have expertise. They know their symptoms, they know the history of their own bodies, they may know which treatments work, and they may be able to impute localized causes for their own disease. Patients may become expert at using medical technology and interpreting readings from blood pressure gauges, blood sugar monitors, and the like. They also have expertise in negotiating treatment plans with doctors and in assessing which doctors are likely to be sympathetic. Lastly, we should not forget that lay people can on occasion play an important role by pointing to new symptoms and causes which doctors may have systematically overlooked, particularly in the work environment.

Nevertheless, this no-doubt legitimate and often unrecognized expertise does not give patients the last word in deliberating over the reality of the disease conditions we have encountered in this chapter. That requires a rather different sort of expertise. Just because patients subjectively know their symptoms and feel they are real makes them no better qualified to pronounce on the complex etiology and epidemiology of disease than a victim of a car accident who pronounces on car safety. Certainly knowing the disease from the inside can give you insights and allows you to sympa-thize with sufferers; it can give you the motivation to learn as much as you can and to proselytize for more research to be done; and it can on occasion lead to medical science changing as it did with the actions of the AIDS activists and in the Lorenzo's oil case; but it is no substitute for epidemiological, pharmacological, and physiological studies.

Therefore, we end this chapter on a cautionary note. Patient advocacy can have a disturbing and counterproductive effect. We have seen legitimate medical experts lambasted, censured, and even silenced simply because they take a position on the reality of a disease which the activists find uncomfortable. Medical experts make mistakes, they have to deal with uncertainty, they may on oc-casion come under the influence of outside pressure from com-mercial interests (but as we have noted, so too do the activists), but, ultimately, they must be permitted to make the major contri-bution to the often highly technical, subtle, and complex debate surrounding the definition of a new disease. Unfortunately in the cases of CFS and fibromyalgia, doctors cannot now say what they really believe; patients sometimes cannot get what might be the best treatment (e.g., psychotherapy) because of their own views about the disease; and lastly society as a whole suffers from the un-necessary undermining of medical expertise and legitimacy. Only if both sides come to terms with the expertise which the other can offer can both benefit from a partnership that will help us under-stand and treat these often cruelly crippling chronic conditions.

Defying Death

Cardiopulmonary Resuscitation (CPR)

It is a Sunday morning in the late fall of 1991. The two authors of this book are sitting in a café in downtown Ithaca sharing a leisurely breakfast. They feel good about life—after a week of intensive work, they have just completed the first draft of what is to become the first volume in this series—*The Golem: What You Should Know about Science*. Suddenly from a nearby table comes an alarming gurgling sound. An elderly woman has collapsed over her plate. We glance at each other—is this serious or is she choking on a crumb? The waitress comes over. "Better call 911," she says to herself as much as to the other diners. The woman continues to gurgle. Pinch's and Collins's eyes meet. "Nothing much we can do I suppose," mutters one of them. Then the other remembers—there is something he can do.

When he was a young lecturer, training by the UK Red Cross had been offered to volunteers by his university department. He had remembered the sad story of a friend forced to stand helpless as someone had collapsed and died in front of him and decided to take up the opportunity. The six-week basic training in first aid and medical emergencies was followed by a test—the assorted collection of academics who had volunteered each administered Cardiopulmonary Resuscitation (CPR) to a dummy under the watchful eyes of the Red Cross officer. The new skills were soon called into play for real. During a departmental dinner a visiting research fellow ate some brazil nuts buried in one of the sauces and alerted her fellow diners that she was starting to experience the symptoms of a severe allergic reaction. The freshly trained medical aid knew he had to call an ambulance and monitor her basic functions until the professionals arrived. She recovered completely and being a "first aid" officer had felt good. That had been five years ago.

The collapsed woman gives another loud gurgle. And our author decides to go into action. He checks the woman's pulse and airway—no obvious obstruction. Her breathing is sporadic. He lays the woman on the floor and begins mouth-to-mouth resuscitation. Should he start heart massage too? The emergency medical workers arrive and thrust an oxygen tube down the woman's throat, give her several shots, and clamp a heart defibrillator across her chest. They administer several large jolts as they carry her to the ambulance.

The drama in which we had participated is becoming part of normal life—especially in the United States and in other advanced industrialized countries. Rather than let someone who collapses die, amateurs and professionals alike immediately try to resuscitate them. Heart defibrillators and oxygen tanks are now available in many public spaces in the United States, such as airports. It is part of modern life that heart attacks and strokes are an ever-present threat and a major, if not *the* major, killer (there are nearly half a million sudden deaths a year in the United States alone). An attack may happen to anyone, anywhere, at any time. The desired response is to give medical attention to the victim as soon as possible. According to the British first aid course attended by the active author, timely intervention is literally a matter of life and death. The sooner you get the lungs working and the heart pumping, the greater the person's chances of survival. In the immediacy of the emergency there was never a question whether this was true or not—current medical opinion said it was. Indeed, all who had attended the first aid course hoped that if they were the unfortunate person to collapse, someone trained in such techniques would be on hand to resuscitate them. Here we begin to question this commonly accepted medical wisdom. Tracing the history of resuscitation and looking at a modern analysis of its effectiveness suggests that things are not so clear.

Resuscitation in History

Resuscitation techniques, like most areas of medical practice, have a lengthy history and have undergone dramatic transformation with the rise of modern medical knowledge. Different methods have come and gone. There has been a series of stuttering changes set in varying contexts of beliefs about death, dying, and human dignity. We will encounter resuscitation methods that are held to be effective even though there is no known physiological basis for them. Conversely, we will encounter methods carefully researched in medical laboratories that have proved ineffective in the field. The world of resuscitation is a world populated not only by the men and women of science and medicine but also by lay people who apply the techniques and act as "their brother's keeper" whenever sudden death threatens. Unsurprisingly, some of the key breakthroughs in this field in the last century have come from the military, where sudden death and its problems are all too familiar. Despite the uncertainties and the changes in methods from era to era, there are two constants—sudden death and people's attempts to overcome it. People have faith that the techniques are effective; they always proceed as if they are. This was true in the eighteenth century and is equally true today.

In medical texts, current resuscitation techniques are often traced back to biblical roots:

32. And when Elisha was come into the house, behold, the child was dead, and he laid upon the bed.

33. He went in therefore, and shut the door upon them twain, and prayed unto the Lord.

34. And he went up, and lay upon the child, and put his mouth upon his mouth, and his eyes upon his eyes, and his hands upon his hands: and he stretched himself upon the child; and the flesh of the child waxed warm.

35. Then he returned, and walked in the house to and fro; and went up, and stretched upon him: and the child opened his eyes." (2 Kings 4:32–35)

When religion held sway, it was only God who could bring back the dead. For humans to try was not only useless but sinful. Over time, however, human intervention has replaced divine intervention. Instead of being the final and irrevocable passage in life's journey, death has slowly become a journey that can be diverted or slowed by humans. To understand the role of resuscitation we must distinguish between "clinical death" and "biological death." Clinical death means failure of circulation, breathing, and the like; biological death is the irreversible atrophying of the human organism. This gap creates the space for resuscitation. The first systematic efforts to create such a gap and fill it with activity seem to have arisen in the eighteenth century.

Death by drowning has always been common and remains so today (only road accidents are a greater cause of accidental death among young people). It is perhaps little surprise that the resuscitation movement started in the two countries where people lived close to the water, the Netherlands and Britain. In 1767, the Dutch founded a society for resuscitating the drowned, which within four years claimed to have saved over 150 people. Seven years later in Britain the Royal Humane Society for the Apparently Dead was formed from the earlier Society for the Recovery of Persons

Apparently Drowned. Its founder and driving force, Dr. William Hawes, reminded his fellow members in 1774 that 125 people had drowned in London the previous year: "Suppose that one in ten were restored, what men would think the designs of this society unimportant, were himself, his relation or his friend—that one?" (quoted in Timmermans, 34). But many people, especially churchmen, objected—it was too like raising the dead, and God was the only one allowed to do that.

To overcome skepticism, the society encouraged people to collect testimonies of successful resuscitations—each case needed three credible witnesses or one learned man such as a priest, physician, or army officer. Because of religious objections, Hawes and his colleagues sharply distinguished between reviving and resurrection: "The former is merely to rekindle the flame of a taper, by gently fanning the ignited wick; the latter to reanimate a corpse, after the vital spark is totally extinct" (quoted in Timmermans, 35). Hence the motto of the society *Lateat Scintillula Forsan* (Possibly a little spark might yet lie hid). Eventually the society won the blessing of the Church, especially with its attempts to interrupt and revive potential suicides (suicide was considered the most evil form of death). One Quaker member pointed to examples in nature, like frozen eels being revived after gentle warming. If God had bestowed simple animals the power to revive, then humans could safely engage in resuscitation efforts as well. Hawes appealed to government officials by ingeniously pointing out that revived murder victims could help solve the crime of which they were a victim! He also played on public fears of live burial. By 1787 the public argument had been won and George II lent the society his patronage.

The society showed an impressive revival rate of 43.7 percent in its early years (out of 1,706 cases). But this rate mixed together cases such as people thrown into the water during a storm screaming to be rescued, and people who had lost consciousness because of smoke inhalation. The term "resuscitation" covered a wide category of rescue situations. Over time it was increasingly recognized that the best results were obtained near the waterfront, and the effort became more and more specifically associated with drowning (Timmermans, 37).

The resuscitation techniques initially used by the Royal Humane Society were those which found favor in the Netherlands and included "warmth, artificial ventilation, administering tobacco smoke rectally or fumigating it, rolling the body over a barrel, rubbing the body, and bleeding from a vein, along with the accessory means of vomiting, sneezing and administering internal stimulants" (Timmermans, 38). What was considered the best method often changed, and the society constantly recommended (and on some occasions forbade) different techniques.

The application of warmth to the victim was a consistently popular technique. It fitted well with Greek physician Galen's theory that warmth was an essential part of vitality. It was also clear that a dead body was a cold body. Although death was marked by loss of breathing, the importance of breathing to successful resuscitation was contested throughout the nineteenth century with arguments pro and con the use of artificial respiration by bellows. Bellows had long been known by anatomists to keep animals alive during experiments, but in 1837 they ran into disfavor after French researchers reported that an animal could be killed by sudden inflation of its lungs and that bellows could produce emphysema (liquid in the lungs) and pneumothorax (escape of air into the chest cavity leading to collapse of the lung) in dead animals. No less a person than Sir Benjamin Brodie, the president of the Royal Humane Society, stated that respiration could not restart a stopped heart. Interestingly, mouth-to-mouth ventilation was experimented with for a short time before being abandoned in 1812 because the exhaled breath was considered poisonous.

In 1857 Dr. Marshall Hall, after realizing that artificial respiration was not in the Humane Society's recommended procedures, carried out a series of experiments on corpses that led to a new theory of death by drowning. He concluded that drowning was similar to anesthesia and poisoning because all involved the exhalation of carbon dioxide. This put a renewed focus upon ventilating the lungs. Hall wanted to avoid the problem of the tongue falling back and occluding the airways as happened when the victim was revived faceup. The solution was to put the victim in a facedown position. Hall essentially turned the old practice of rolling the body over a barrel into what he now called the "postural method" of artificial respiration. Expiration was produced in the facedown position by applying pressure on the back over the thorax and abdomen. Inspiration was produced the moment the pressure was withdrawn and was completed by rolling the patient on the side. A young surgeon, Henry Silvester, at the same time proposed another important manual ventilation technique. Rather then looking at what failed in death (as Hall had done), Silvester tried to imitate the natural respiratory movements of a living person. He preferred the faceup posture because he believed this allowed the user to check for airway blockage. The rescuer stood at the victim's head, grabbed the elbows, and pulled the arms back to the ears to stimulate inspiration. To stimulate expiration, the arms were brought back and pressed upon the chest.

The new techniques of Hall and Silvester were both supported by what at the time were considered to be sound theories, research results, and an impressive degree of success. The Royal Humane Society was now presented with a dilemma. Which technique should it adopt? A comparative study carried out on cadavers showed the Silvester method to be superior for ventilating the lungs, but not everyone was convinced.

In 1889 a new Royal Humane Society president, Edward Schafer, initiated another review and proposed yet another artificial ventilation technique based upon applying intermittent pressure on the thorax—again with the patient in a facedown position. Debate in the society now raged over the merits of the Silvester method versus the Schafer method. Silvester himself opposed the facedown method because the posture of the operator "athwart the patient" in respect to the female patient was "undesirable" (Timmermans, 42). While earlier researchers had experimented on warmed cadavers or dogs, modern techniques devised by Schafer used human volunteers who repressed the breathing reflex— breathing output was measured with displaced tidal air volume (the amount of air displaced in a "tidal cycle" of inhalation and exhalation). The tidal volumes, however, from ten different methods employed on five volunteers were inconclusive.

In 1909 one reviewer of the society's resuscitation records noted that "every resuscitation technique—forbidden or recommended, physiologically sound or far-fetched, with or without artificial respiration—seemed to be able to save an impressive number of human lives" (quoted in Timmermans, 41). One unusual method was the technique of tongue traction introduced by the Frenchman Laborde in 1892. The method consisted of "opening the mouth and pulling out the tongue with some degree of force" (Timmermans, 4). Because there were no physiological grounds for this technique it was placed on the list of forbidden methods, yet in France Laborde noted its successful use in sixty-three cases.

Eventually society members agreed that warming the body and artificial ventilation were the best means for reviving victims, and the terms "artificial respiration" and "resuscitation" started to be used interchangeably. For the first half of the twentieth century artificial respiration with either the Schafer prone-pressure method or the Silvester technique became the standard. The Schafer method was most popular in Britain, France, Belgium, and the United States, and the Silvester method had supporters in Germany, Holland, and Russia. Exhalation became the key vital

139

sign to look for in terms of need for resuscitation. Death resulted from lack of oxygen in the lungs. A small mirror placed against the victim's mouth was effective for determining exhalation. If the mirror fogged up, the victim was alive and did not need to be resuscitated, only warmed. If the mirror didn't fog up, artificial respiration had to be started immediately. The mirror was in effect one of the first portable diagnostic tools to be used in resuscitation. Boy Scouts in the UK in the 1950s trained to "be prepared" by carrying such mirrors.

After the Second World War, research on resuscitation moved to the United States. A review of U.S. case studies gathered from the Coast Guard and the fire departments of Chicago, Detroit, and Los Angeles revealed the dominance of the Schafer method, with a survival rate of 6.7 percent. Despite its widespread use, the Schafer technique had not done well in the war, where numerous soldiers on troop supply ships had drowned. The war provided an additional spur to new research because it was feared the Germans might use nerve gases capable of paralyzing the respiratory muscles. In 1948 the U.S. Army and the National Research Council organized a conference of physicians to compare all the different methods. Conference participants agreed that they lacked data to choose the best method, so extensive new comparative experiments were initiated. Amazingly it was discovered that the Schafer technique, which for fifty years had been the dominant resuscitation technique and had apparently saved thousands of lives, was worthless on experimental grounds. It was found that it could not move dead air in the trachea, meaning that no fresh oxygenated air could enter the lungs. The Silvester method was similarly found wanting because the patient was kept in a faceup position, thereby contributing to airway occlusion by the tongue. A new method of artificial respiration, known as the "back-pressure arm-lift" method, was adopted. It was introduced at a conference held in 1951 attended by representatives of the American National

Red Cross, armed forces, Boy Scouts of America, AT&T, Bureau of Mines, Campfire Girls, Girl Scouts of the United States, YMCA, AMA, and numerous public utility and civil-defense organizations. A two-page standard was published and training films were issued, all as part of a widespread publicity campaign. It at last seemed that postwar research had identified the best method of resuscitation. A standard had been agreed upon and had been taken up by numerous organizations and was put to daily use in the attempt to save lives.

Only four years passed before the new standard was in trouble. Research on the problem of resuscitating young children carried out by Harold Rickard, a captain in the U.S. Navy and a self-proclaimed practitioner of resuscitation for thirty-five years, drew attention to the problem of airway obstruction. Rickard knew from his own practical experience that all the recommended techniques were useless because the relaxed tongue of the victim obstructed the airway. Inspired by Rickard (who lacked any clinical backing for his ideas) an anesthesiologist, Peter Safar, used x-rays and results from spirometers (devices for measuring air flow) to show that all manual ventilation methods, regardless of whether the victim is faceup or facedown, suffered the same problem. This finding again was astonishing. It seems that all the laboratory-derived measurements had used intubation (the insertion of a tube), and this had prevented the obstruction occurring. The tube used for measuring the airflow had pushed the tongue out of the way! Although the manual ventilation techniques could be made to work by hyperextending the neck and placing the victim in the faceup position, it was clear that Safar's work lent encouragement to an old technique that was receiving new attention—mouth-to-mouth resuscitation.

The military again played an important role in the new development. In 1950 Dick Johns and David Cooper, as part of an army research group, developed mask-to-mask resuscitation for nerve gas casualties in a contaminated environment. Bemoaning the "stupidity" of the U.S. Army and its approach to manual ventilation, Cooper and Johns devised a way of linking two gas masks together so that the exhaled breath of the rescuer could enter the lungs of the victim. They experimented with the device on each other and on some dogs and wrote a report. They tried to get the U.S. Navy interested in their device, but to no avail. The report, however, caught the eye of a young physician, James Elam, who had instinctively turned to mouth-to-mouth ventilation as the only way of keeping iron lung polio victims alive during power failures. In 1950, in his first university position, Elam researched mouth-to-mouth ventilation by blowing into the tracheal tube of postoperative patients still under anesthesia with ether. At the same time an assistant drew blood to measure the oxygen content. He found results vastly superior to those obtained using manual methods. He was invited to attend the famous 1951 conference where the new manual ventilation standard was proclaimed. Tacked on at a special session at the end of the conference, Elam was eager to detonate his "bomb." But it fizzled when the leading researchers showed little interest in a technique that they thought of as little more than "common sense" (Timmermans, 48).

Elam tried to proselytize his new technique in Washington and published articles on it in leading medical journals to little effect. His breakthrough came in 1956 when he accepted a ride with Safar from an anesthesiology conference they both were attending in Kansas. Safar, who had become one of the leading U.S. authorities on resuscitation techniques, was now chief anesthesiologist of Baltimore City Hospitals and had started to experiment with mouth-to-tracheal tube ventilation to inflate patients' lungs to ascertain bilateral chest movements. He and Elam compared both methods and showed decisively the merits of mouth-to-mouth ventilation. Soon other leading researchers had confirmed the results, and in 1960 a group of international researchers tested mouth-to-mouth resuscitation in six metropolitan areas in more than 1,000 anesthetized patients and recommended that mouth-to-mouth become the only method to resuscitate everyone except newborns. They strongly recommended that it be taught to professionals and laypeople. Commercial companies tried to promote a more complicated form of the technique with an oral airway to be inserted into the victim's throat, but the team of researchers resisted, insisting that the new technique be simple, safe, and easy to learn.

Although Safar and Elam had the research community on their side, they were well aware that theoretical arguments had earlier been used to justify new manual techniques such as the facedown position. In introducing the new technique, Safar and Elam drew attention to the accessibility and visibility of the victim's face to check the airway and to apply mouth-to-mouth resuscitation. These practical arguments won out, and the mouth-to-mouth technique is the standard used in CPR training to this day.

External Heart Massage

Just when it seemed that the quest for the best resuscitation technique was complete, research again shifted dramatically. External heart massage was first developed at the medical laboratories of Johns Hopkins University. William Kouwenhoven, an engineer, had been asked to develop a portable defibrillator for use in the electrical utility industry where over 50 percent of workers suffering severe electric shocks died from ventricular fibrillation (the heart going out of rhythm). It was a graduate research assistant, Guy Knickerbocker, who first showed that chest compressions could raise blood pressure and help cardiac arrest victims. In July 1958 he was experimenting on a dog with the heavy, fifteen-pound paddles of the portable defibrillator then in use, when he noticed an increase in blood pressure. With the aid of a neighboring laboratory researcher, he managed to keep another dog with

143

cardiac arrest alive for eight minutes by means of heart compressions. Over the next year Knickerbocker and Kouwenhoven tested and refined their new technique, which they called "external chest massage." They showed that up to five minutes of ventricular fibrillation could be overcome using external massage. Since Kouwenhoven's goal was to equip every truck used in the electricity industry with a portable defibrillator, this result was significant, because it meant that maybe not every truck would need one. The finding was sufficiently impressive for the head of the laboratory, the world-renowned surgeon Dr. Alfred Blalock, to assign Dr. James Jude, a resident in surgery, to the project to provide medical legitimacy.

Jude immediately found an important new use for the technique. Sometimes cardiac arrest occurred during surgery as an unintended consequence of the anesthetic. The only solution then available was to quickly open the chest of the patient and massage the heart by hand. Surgeons even used to carry an extra scalpel in their breast pockets for this purpose. Such surgery always caused complications, usually serious infections. Jude had the chance to try the new technique when a female patient admitted for a gallbladder operation unexpectedly suffered a cardiac arrest. Intubation failed, and when Jude saw the blood pressure and pulse disappear, he put his hands on her chest and started external cardiac message. After a dramatic two minutes of massage a pulse developed and a spontaneous but shallow respiration. The patient eventually recovered fully, and no use of artificial respiration was made. After four more successful interventions with patients, Jude, together with Knickerbocker and Kouwenhoven (who had continued to carry out laboratory tests), published an article in the *Journal of the American Medical Association*. They famously wrote, "Anyone, anywhere can now initiate cardiac resuscitative procedures. All that is needed are two hands" (quoted in Timmermans, 52). These words signified one of the most important changes ever in resuscitation research. As well as marking a shift to cardiac resuscitation—before, it had been assumed that to revive a patient meant reviving first the lung function—it also signaled the universal import of resuscitation. Anyone could do it anywhere on any sort of patient—not only drowning victims, not only the apparently dead under certain circumstances, but any dying person, including those who previously would have been thought beyond hope. It meant a new clinical definition of the dying process—no longer was absence of pulse the sole criterion, since a pulse could be restarted. As Stefan Timmermans writes, "When mouth-to-mouth ventilation was combined with chest compressions to form CPR . . . at a conference in Maryland in 1960, sudden death was just one more roadblock waiting to be cleared in modern medicine" (Timmermans, 53).

CPR for All

But how effective was CPR? Physicians initially opposed the transfer of heart massage "rights"—a specialist medical technique—to lay people. If the compressions were not given correctly, internal damage to delicate organs could easily arise. To overcome such objections training programs had to be initiated. Also it was clear that to be effective the administering of CPR would have to be followed up with other medical interventions and technologies such as injections of cardiac drugs, oxygen, and defibrillation. Also a system of emergency medical response and rapid transportation to hospital was needed.

It was not until 1973 that such a full system was put in place in the United States. A National Conference on CPR and Emergency Cardiac Care recommended integrating universal CPR with emergency paramedic-based ambulances. The new system drew a distinction between *basic* and *advanced* lifesaving interventions. The basic system was CPR to be taught to everyone from eighth grade upward, although initially groups with the greatest needs like police officers, firefighters, rescue workers, lifeguards, and so on, received priority. Advanced cardiac life support, such as intravenous fluid lines, drug injection, defibrillation, and cardiac monitoring, was reserved for specially trained health-care professionals. To implement the basic training, the American Heart Association standardized and disseminated CPR protocols and organized the training and certification of instructors. Coronary care in American hospitals was reorganized to mesh with the new CPR-revived patients dropped on their doorsteps in droves by ambulances (now also equipped with CPR technologies).

In view of the massive investment of human and material resources in CPR and the associated reorganization of the emergency care system, it is interesting to ask what sort of survival rate was expected. Interestingly the thirty-two-page final document from the 1973 conference did not predict how many lives would be saved. This omission was not deliberate. Swept away by a tide of optimism, the conference organizers simply assumed that with public CPR and an emergency medical system in place, a significant number of lives would be saved.

Assessing survival rates is not easy. To this day a general comprehensive CPR survival rate for the United States is unavailable. Without national databases, medical researchers and policymakers do not know how many people have undergone CPR. Researchers have had to rely on regional survival rates based on short-term, small-scale studies. Such figures tend to vary enormously. One comparison of survival rates in twenty-nine U.S. cities and abroad between 1967 and 1988 found a diversity of rates, ranging from 2 percent in Iowa to 26 percent in King County, Washington. Taken together these survival rates contrast starkly with the optimism of 1973. The studies also confirmed, however, that universal CPR was well established. The general public had indeed learned about CPR, and in several communities over half the resuscitative efforts began with bystanders. In addition, most emergency medical systems had been successfully re-organized.

Researchers began to worry about what led to such a variety in survival rates. A number of 2 percent in Iowa meant that only one in fifty resuscitative efforts was successful, whereas in Seattle one in four efforts seemed to save a life. But to make sense of these numbers, resuscitation in Seattle and Iowa had to start from the same baseline. For instance, take just one variable known to affect chances of survival—the response time. This is the time between collapse and initiation of the resuscitation efforts. There is no clear definition of "response time" in the literature. As one reviewer, Dr. Mickey Eisenberg, found, it may involve all or some of the following: recognition, decision to call, calling, dispatch interview, dispatching, travel from station to scene, and travel from scene to patient's side (Eisenberg, quoted in Timmermans, 70). In addition to response time, Eisenberg found variations in definitions for such basic terms as "cardiac arrest," "bystander CPR," "witnessed arrest," "ventricular fibrillation," and "admission." Even more significant for the purposes of a meaningful comparison, researchers defined the two key components of a survival rate—resuscitation and survival—differently. Some defined any attempt at CPR as a resuscitative effort, while others restricted it to particular cardiac rhythms such as the rapid heartbeat known as ventricular tachycardia. "Survival" was also ambiguous. For some studies it meant discharge from hospital with minimal neurological damage; for others it referred to admission to the intensive care-unit with a viable pulse. If we add in the inevitable differences between regional medical systems, then it can be seen that the survival rates are even harder to interpret.

In order to overcome these endemic definitional problems, the Utstein Consensus Conference was held at Utstein Abbey on a small island off the Norwegian coast in 1990. This conference standardized definitions and proposed a uniform formula for the

147

calculation of the survival rate. The new formula was "the number [of patients] discharged alive [from the hospital] divided by the number of persons with witnessed cardiac arrest, in ventricular fibrillation, of cardiac etiology" (Timmermans, 73). In other words, the victim must be suffering from an underlying heart condition and not one caused by other events such as drowning or electric shock.

This definition of survival rate was much narrower than had been employed previously. It ruled out many of the incidents and accidents (including drowning) which had counted as resuscitations over the previous two centuries. Also excluded were unwitnessed cardiac arrests and patients on whom bystanders did not initially perform CPR. Because only cases with the best chances of survival were included, the survival rate was inflated when compared with studies using a more comprehensive definition. Indeed, most conditions prompting CPR (60–80 percent of all cases) were now excluded from the statistics. On the other hand, the requirement that the patient be discharged alive from the hospital created a relatively high standard for survival (although discharge standards varied widely throughout U.S. hospitals and around the world).

Since 1991 a few studies have been carried out using the Utstein standards. Although the healthiest and most homogeneous sample possible was examined, the survival rates were disappointingly low and still varied greatly. For example, a study in Chicago found a survival rate of 0.8 percent for African Americans and 2.6 percent for Caucasians. The article in which these results were published was aptly titled "Outcome of CPR in a Large Metropolitan Area—Where Are the Survivors?" In New York City the survival rate was also low at 1.4 percent. On the other hand Oakland County, Michigan, had a survival rate of 14.9 percent. The medical literature that interpreted these figures was, as Timmermans notes (74), persistently optimistic. The lower rates were put down to the poorer medical services in large metropolitan areas. With a mature emergency system and the necessary political will, the supporters of CPR argued, survival rates as high as the 30 percent obtained in Seattle ought to be achievable everywhere. Early defibrillation was now seen as the key to improving survival rates, with some studies indicating that 80–90 percent of all survivors had been treated for ventricular fibrillation—a technique not available to the unassisted bystander. The current strategy in the United States is thus to promote the widespread availability of defibrillators such that they become standard equipment, like fire extinguishers, in airplanes, health clubs, offices, and the like.

Statistics on rates of survival due to defibrillation are little better than other statistics in this field, however, varying from area to area and showing little convincing evidence of a breakthrough in survival rates. Perhaps the more profound point about the statistics is made by Timmermans. He argues that since the medical community and the general public has invested so much belief in CPR, no statistics, however low, could damage the idea of its efficacy. Poor rates of survival are always interpreted as a sign that emergency services and the medical infrastructure need to be improved in order to obtain better rates. The underlying effectiveness and need for CPR are rarely questioned. Even the American Heart Association, a booster for CPR, acknowledged in 1991 that the current estimate of the numbers of people who live to be discharged from hospital after suffering cardiac arrests was no more than 1–3 percent, and that because of poor data the true percentage was "probably even less" (Timmermans, 4).

Lastly, it is worth asking what it means to survive CPR. According to the Utstein criteria, survival means discharge from hospital. But this leaves unanswered the questions of what state such patients are in, what happens to them afterward, and what happens to patients who undergo CPR but who are not discharged and continue to live severely impaired lives, perhaps in long-term nursing

149

care or a vegetative state? There are a few studies of this aspect of
the problem, and Timmermans summarizes the situation as follows: "These studies provide startling and dramatic findings. Survival rates hide the Russian roulette aspect of resuscitative efforts.
The term 'survival rate' highlights lifesaving while it glosses over
the possibility—indeed, the likelihood—that the same interventions create neurological impairment. Because different vital organs can be restored after varying time spans, hearts and lungs
will inevitably be restored while brains will not. With CPR, we save
lives but we produce people with a range of disabilities" (Timmermans, 81).

By focusing on survivor rates we forget that most people who
undergo CPR may not survive in the way we would want them to.
The dilemmas are the same as for anyone who has gone through
the heart-wrenching process of deciding what to do or not to do for
a family member made comatose by a severe stroke. Survival rates
also distract from the underlying truth, which is that most people
who undergo CPR actually die—Timmermans asks pertinently
why the statistics are presented as "survival rates" instead of as
"mortality rates"?

How aware is the public of these figures? It seems that most
people undergoing CPR training as our author did in the mid
1980s have no idea how low the survival rate actually is. TV does
not help. In 1996 researchers analyzed the portrayal of resuscitation on three popular U.S. TV shows, *ER*, *Chicago Hope*, and *Rescue 911*. The resuscitation rates were unrealistically high—immediate survival was an incredible 75 percent and longtime survival
was 67 percent. In addition most cardiac arrests were due to
trauma and figured children, teenagers or young adults (the average age of cardiac arrest in Seattle is 65 years), and the shows focused on miraculous recoveries. When questioned about this, the
producers justified the misrepresentations on the grounds that it
might encourage young people to learn CPR. And perceptions of

CPR are of course quite crucial to the whole process. If the effectiveness of CPR as a technique is doubted, then fewer people might learn it and thus reinforce its very lack of effectiveness.

In view of the surprising conclusion—that resuscitation rates are variable and are disappointingly low for most large metropolitan areas and that the quality of life of the person who survives may be severely impaired—we must ask why we continue to believe in the efficacy of CPR and devote so many of our resources to it? The answer lies in a combination of faith in modern medicine with all its attendant uncertainties and our attitudes toward death and the dying. In this case we need medicine to give us the hope of escaping from death's dark door, even if, realistically, the numbers it can help are very few. The effectiveness of CPR, as Stefan Timmermans has written, is a "revered cultural myth perpetuated by 'real-life' television shows and the organizations promoting CPR. The technique spins a tale of medical heroism, medical magic to overcome the adversity to death, and the holy grail of a prolonged life in everyone's reach" (Timmermans, 5).

Timmermans believes that his careful analysis has shown that cardiopulmonary resuscitation (CPR) is ineffective, and his conclusion is that the resources spent on it should go elsewhere. CPR could be continued under limited circumstances as a "passing ritual," allowing relatives a little more time at the bedside of the dying person. We have noted that at least one study of CPR suggested it had a 30 percent success rate even though the bulk of the analysis found success rates of one or two percent. The one optimistic study does seem to invite more research. But let us accept for a moment that Timmermans's interpretation of his statistics is right and that the cost of widespread provision of CPR training and equipment is an inefficient use of resources. In that case, it is hard to disagree that further public spending on equipment might be misguided if the money could be redirected somewhere more beneficial. On the other hand, were CPR training, of the kind undergone by our author, to continue on a relatively inexpensive voluntary basis, so that individuals knew there was something they could do to try to help a person in trouble, there seems no reason to deny that hope. And, of course, as we have argued throughout, even a 1 or 2 percent improved chance of success amounts to 100 percent for those one or two in one hundred; that is the argument for medicine as succor. The economics of scarcity, as we have said, is the constant companion of medicine as a collective enterprise: if it is statistically robust, what Timmermans's study shows is that CPR should be cheaper rather than more expensive—it does not show it should be stopped. We do not know what happened to the lady in the Ithaca café, but to be able to offer some hope for what was in this case, little cost, still seems to be the right thing to have done.

Postscript (August 2004)

The very latest data show no improvement in overall survival rates for those given CPR. In a recent article published in the *New England Journal of Medicine,* the authors write, "Most communities have overall survival rates of less than 5% for cardiac arrests occurring outside the hospital. There is no evidence of these rates increasing, despite extensive use of advanced treatments and technology."[1]

The AIDS Activists

7

This chapter, with the exception of this short introduction, is a reprint of a chapter from *The Golem at Large*.[1] We reproduce the AIDS story here because it is so relevant to the themes of this book. The case shows, first, how difficult it is to conduct randomized control trials in the ideal way required by the theory of statistical analysis; the "gold" from which the gold standard is made shows a disturbing tendency to tarnish. Second, the theme of the tension between the best available standards of scientific testing and the needs of the individual for succor is poignantly illustrated by the sharing of drugs between placebo and treatment groups, which is the most dramatic moment in the following account. Third, the way that unqualified groups can acquire interactional expertise and even a modicum of contributory expertise in an esoteric science is beautifully illustrated by this case. The fourth lesson is that the acquisition of such expertise is very hard work and should not be taken on lightly; for the AIDS activists to win acceptance from the scientific community they had to participate in the discourse of science, not just master the vocabulary or read the literature. The final warning lesson for those who consider that it is a trivial matter to acquire medical expertise is that when the activists did master enough of the science to speak on equal terms with the scientists, they found that what the scientists were saying made more sense than they had first thought! Unsurprisingly, other groups of activists, who had not undergone such a complete passage of scientific socialization, believed their colleagues had "gone native," allowing themselves to be coopted. That is a tension familiar in the participatory social sciences.

Acting Up: Aids Cures and Lay Expertise

On April 24, 1984, Margaret Heckler, U.S. Secretary of Health and Human Services, announced with great gusto at a Washington press conference that the cause of AIDS had been found. A special sort of virus—a retrovirus—later labeled as HIV, was the culprit. Vaccinations would be available within two years. Modern medical science had triumphed.

Next summer, movie star Rock Hudson died of AIDS. The gay community had lived and died with the disease for the previous four years. Now that the cause of AIDS had been found and scientists were starting to talk about cures, the afflicted became increasingly anxious as to when such cures would become available. Added urgency arose from the very course of the disease. The HIV blood test meant that lots of seemingly healthy people were facing an uncertain future. Was it more beneficial to start long-term therapy immediately or wait until symptoms appeared? Given the rapid advance in medical knowledge about AIDS and the remaining uncertainties (even the cause of AIDS was a matter of scientific debate), was it better to act now with crude therapies or wait for the more refined treatments promised later?

AIDS the "Gay Plague"

AIDS is not confined to homosexuals, but in the United States it was first described in the media as the "gay plague," and gays as a community were quick to respond to its consequences. The gay community in the United States is no ordinary group. The successful campaigns for gay rights in the sixties and seventies left them savvy, streetwise and well organized. The large numbers of well-educated, white, middle-class, gay men added to the group's influence. Although Main Street America might still be homophobic, there were sizable gay communities in several big cities, with their own institutions, elected officials, and other trappings of political self-awareness.

Being gay had become relatively more legitimate, to some extent replacing earlier views of homosexuality which treated it as a disease or deviant condition. AIDS now threatened to turn the clock back and stigmatize gay men once more. In the public's mind, the disease was a biblical judgment brought on by gays' reckless promiscuity; with Reagan in power, and the right wing in ascendancy, AIDS became a way to voice many prejudices. For instance, conservative commentator William F. Buckley Jr. proposed in a notorious *New York Times* op-ed piece in 1985 that "everyone detected with AIDS should be tattooed in the upper forearm to protect common-needle users, and on the buttocks, to prevent the victimization of other homosexuals" (Epstein, 187).

The charged atmosphere was evident in early community meetings held in the Castro district, at the heart of San Francisco's gay community. Captured movingly in the film *And the Band Played On* (based upon Randy Shilts's book of the same name), the gay community agonized over the difficult decision to close the bathhouses—one of the most potent symbols of 1970s-style gay liberation. AIDS struck deep at the core institutions and values of the newly emancipated.

Grassroots activist organizations soon arose, dedicated to acquiring information about AIDS and how to fight it. Those who tested positive for HIV were told to expect several years of normal life before the onset of the disease. AIDS activism not only fit the psycho-

155

logical, physical, and political circumstances in which people found themselves, but also, unlike many forms of activism, it promised direct benefits in terms of better medications and, perhaps, a cure.

The gay community was predisposed to be skeptical toward the world of science and medicine, especially as homosexuality had for years been labeled as a medical condition. To intervene in the AIDS arena the community would have to deal with some very powerful scientific and medical institutions. As we shall see, AIDS activists turned out to be remarkably effective in gaining and spreading information about the disease and its treatment. They also contributed to the scientific and medical debate to such an extent that they came to play a role in setting the AIDS research agenda and on occasion did their own research. How such expertise was acquired by a lay group and applied so effectively is a remarkable story.

We tell this story in two parts. In part 1 we trace some of the science of AIDS and document historically how the activists increasingly came to play a role in AIDS research. Part 1 concludes with the first official approval for use of an AIDS drug where the research was conducted to a significant extent by lay experts. In Part 2 we further document the successes of the activists, focusing on one particularly influential group—ACT UP. We show how the lay activists acquired and refined their expertise so that they were able to mount a thoroughgoing political and scientific critique of the way that AIDS clinical trials were conducted. This critique by and large became accepted by the medical establishment.

Part 1

A VACCINE IN TWO YEARS?

From the earliest moments of the AIDS epidemic it was apparent that misinformation was widespread. There were moral panics among the public about how AIDS could be transmitted. More significantly, early public pronouncements about the prospects for a cure were exaggerated. Scientists who had sat in on Heckler's press conference winced when she talked about a vaccine being available within two years. At that point reasonably effective vaccines had been developed for only a dozen viral illnesses, and the most recent, against hepatitis B, had taken almost a decade to bring to market. Dr. Anthony Fauci, head of the National Institute of Allergy and Infectious Diseases (NIAID), was more circumspect when he told the *New York Times,* a few days after Heckler's announcement, "To be perfectly honest . . . we don't have any idea how long it's going to take to develop a vaccine, if indeed we will be able to develop a vaccine" (quoted in Epstein, 182).

Viruses invade the genetic code of the cell's nucleus—DNA—transforming each infected cell into a site for the production of more virus. Viruses, in effect, become part of each of the body's own cells. This is quite unlike bacteria—cell-sized foreign bodies more easily identifiable and treatable by drugs such as antibiotics. To eliminate a virus requires that every infected cell be destroyed without harming the healthy cells at the same time. Worse, in the continual process of virus reproduction it is likely that genetic mutations will arise; this makes a viral disease even harder to treat.

THE PROMISE OF ANTIVIRAL DRUGS

HIV differs from normal viruses in that it is a "retrovirus." It is composed of RNA (ribonucleic acid) rather then DNA (deoxyribonucleic acid). Normally viruses work by turning cells into virus factories using the blueprint of the original viral DNA. The virus's DNA is copied into RNA, which is then used to assemble the proteins forming the new viruses. When retroviruses were first discovered they presented a problem. If they were made solely of RNA, how did they replicate? The answer was found in an enzyme known as "reverse transcriptase" which ensured that the RNA could be copied into DNA. The presence of this enzyme offered the first opportunities for a cure. If you could find an antiviral agent that eliminated reverse transcriptase, then perhaps HIV could be stopped in its tracks. Some antiviral drugs showed early promise in killing HIV *in vitro* (literally, "in glass," that is outside the body, in a test tube).

Infection with the HIV virus can be diagnosed by a blood test—the infected person being known as "HIV-positive." Symptoms of the disease AIDS may not appear until many years later. "Full-blown AIDS" consists of a variety of opportunistic infective diseases which develop because the body's immune system can no longer combat them. "Helper T cells," which are crucial in fighting such opportunistic diseases, become depleted. Even if HIV could be killed *in vivo* ("in life," inside the body), a cure for AIDS might not necessarily follow. Perhaps long-term damage to T cells would have already occurred at an earlier stage of infection. Perhaps HIV infection interfered with autoimmune responses by some unknown route, which meant that the immune system as a whole lost its ability to tell the difference between body cells and invaders.

In any event the path to a cure was likely to be long. An antiviral compound had to be found which could be administered to humans in safe and clinically effective doses without producing damaging side effects. Its effectiveness had to be confirmed in controlled clinical trials with large numbers of patients. Lastly, it had to meet legal approval before being offered for widespread use.

CLINICAL CONTROLLED TRIALS AND THE FDA

Since the thalidomide scandal (thalidomide was developed to treat morning sickness in pregnant women and later was found unexpectedly to cause severe birth defects), the Food and Drug Administration (FDA) required that new drugs be extensively tested before approval. Three phases of randomized control testing were demanded. In phase 1 a small trial was necessary to determine toxicity and an effective dosage. In phase 2 a larger, longer trial was carried out to determine efficacy. Phase 3 required an even larger trial to compare efficacy with other treatments. This process was costly and time-consuming—typically it could take a new drug six to eight years to clear all the hurdles.

In October 1984 the first International Conference on AIDS took place in Atlanta, Georgia. These conferences, attended not only by scientists and doctors but also by gay activists and media personalities, became annual milestones. It was reported that small-scale trials had begun with six promising antiviral drugs, including one called "ribavirin." But phase 1 trials were still a long way off. "We have a long way to go before AIDS is preventable or treatable," Dr. Martin Hirsch of Massachusetts General Hospital concluded in reviewing the conference, "but the first steps have been taken, and we are on our way" (quoted in Epstein, 186).

BUYERS' CLUBS

People dying of AIDS, and their supporters, were impatient with this kind of caution. They were desperate to try anything, however unproven, to halt the march of the deadly disease, and they soon began to take matters into their own hands. Ribavirin was reported to be available at two dollars a box in Mexico. Soon this and other antiviral drugs were being smuggled over the border into the United States for widespread resale to AIDS sufferers. Illicit "buyers' clubs" started to flourish. Wealthy gay patients became "AIDS exiles," moving to Paris, where another antiviral agent, not approved in the United States, was available.

Embarrassed by media stories about "AIDS exiles" such as Rock Hudson, the FDA announced that new antiviral drugs under test would be made available under a long-standing rule for "compassionate use." This meant that doctors could request experimental drugs for their terminally ill patients as a matter of last recourse.

PROJECT INFORM

The San Francisco gay community was a focal point for activism. Project Inform, a leading activist research group, was founded by Bay area business consultant, former seminary student, and ribavirin smuggler Martin Delaney. The aim was to assess the benefits to be gained from new experimental drugs: "'No matter what the medical authorities say, people are using these drugs,' Delaney told reporters skeptical of the idea of community-based research. 'What we want to do is provide a safe, monitored environment to learn what effects they are having'" (quoted in Epstein, 189). Although Delaney had no scientific background, he had personal familiarity with one of the key issues in the upcoming debate: Who assumes the risk a patient faces in taking experimental drugs, the patient or the doctor?

Delaney had previously participated in an experimental trial of a new drug to treat hepatitis. The drug had worked for him, but side effects had led to damage to the nerves in his feet. The trial was terminated and the treatment was never approved because it was thought the drug was too toxic. Delaney, however, considered it a "fair bargain" (Epstein 189), since his hepatitis had been cured.

The prevailing trend in U.S. clinical trials was to protect patients from harm. In 1974, Congress had created the National Commission for the Protection of Human Subjects with strict guidelines for research practices. This had been in response to a number of scandals where patients had been unknowingly subject to experimentation. The most notorious was the Tuskegee syphilis study, where, for years, poor black sharecroppers had been denied treatment so researchers could monitor the "natural" course of the disease.

Delaney, in pushing for patients to be given the right to do themselves potential harm with experimental treatments, seemed to be taking a backward step.

THE TRIALS OF AZT

The efforts of the activists to get more patients into experimental drug treatment programs reached a peak in 1985 when, finally, it seemed that a promising antiviral agent had been found. AZT (azidothymidine) had been developed originally to fight cancer. In this use the drug had failed. It had sat for years on the shelf at Burroughs Wellcome, the North Carolina—based subsidiary of the British pharmaceutical company, Wellcome. In late 1984, the National Cancer Institute (NCI) had approached leading pharmaceutical companies to send them any drugs that had the potential to inhibit a retrovirus, and AZT was dusted off. In February 1985 AZT was found to be a reverse transcriptase inhibitor with strong antiviral activity. A phase 1 trial was immediately carried out. The six-week study of nineteen patients showed that AZT kept the virus from replicating in fifteen of the patients, boosted their T cell counts, and helped relieve some of their symptoms. In transcribing the virus's RNA into DNA, AZT appeared to fool the reverse transcriptase into using it, in place of the nucleoside it imitated. Once AZT was added to the growing DNA chain, reverse transcriptase simply ceased to work and the virus stopped replicating. The problem with AZT was that, since it stopped DNA synthesis in the virus, there was every reason to believe that it might have harmful effects on DNA in healthy cells.

The NCI researchers were cautious in reporting their results because of the placebo effect. Perhaps the effects reported by the NCI researchers were artifacts produced by patients' knowledge and expectation that AZT would help cure them? Although noting the immune and clinical responses to the drug, they warned about the possibility of a strong placebo effect. NCI called for a long-term, double-blind, controlled placebo study to be carried out in order to better assess the potential of AZT.

Funded by Burroughs Wellcome, plans went ahead to conduct this new trial at a number of locations. At this time the testing of new AIDS drugs became more complicated because, with the help of $100 million in funding, NIAID started to set up its own network of centers to evaluate and test a variety of putative new AIDS drugs, including AZT. All this was done under the leadership of

NIAID's head, Anthony Fauci. Creating the new centers took some time, as a whole series of new research proposals and principal investigators had to be vetted. Time was something most AIDS patients did not have.

AIDS activist John James founded a San Francisco newsletter called *AIDS Treatment News,* which went on to become the most important AIDS activist publication in the United States. James was a former computer programmer with no formal medical or scientific training.

In the third issue of *AIDS Treatment News* James reported that large-scale studies of AZT were still months away and that even if all went well, it would take another two years before doctors would be able to prescribe AZT. He estimated that at a death rate of ten thousand a year, which was expected to double every year, a two-year delay would mean that three quarters of the deaths which ever would occur from the epidemic would have occurred, yet would have been preventable.

In James's view a new task faced gay activists and AIDS organizations:

> So far, community-based AIDS organizations have been uninvolved in treatment issues, and have seldom followed what is going on. . . . With independent information and analysis, we can bring specific pressure to bear to get experimental treatments handled properly. So far, there has been little pressure because *we have relied on experts* to interpret for us what is going on. They tell us what will not rock the boat. The companies who want their profits, the bureaucrats who want their turf, and the doctors who want to avoid making waves have all been at the table. The person with AIDS who want their lives must be there too. (Epstein, 195; emphasis added)

James did not regard AIDS researchers as incompetent or evil. It was rather that they were too bound up in their own specialties and too dependent on bureaucratized sources of funding to be able to present and communicate a full and objective picture of what was going on. James believed that lay activists could become experts themselves: "Non-scientists can fairly easily grasp treatment-research issues; these don't require an extensive background in biology or medicine" (quoted in Epstein, 196). As we shall see, James's optimism was not entirely misplaced.

Meanwhile the phase 2 testing of AZT started. On September 20, 1986, a major study made headline news when it was ended early. AZT proved so effective that it was considered unethical to deny the drug to the placebo-taking control group. Dr. Robert Windom, the assistant secretary for health, told reporters that AZT "holds great promise for prolonging life for certain patients with AIDS" (quoted in Epstein, 198), and he urged the FDA to consider licensing AZT as expeditiously as possible. With FDA and NIH support, Burroughs Wellcome announced that it would supply AZT free of charge to AIDS patients who had suffered the most deadly infectious disease, a particularly virulent form of pneumonia (PCP), during the past 120 days. Under pressure from AIDS patients and doctors who considered the criteria too arbitrary, this program was expanded to include any of the 7,000 patients who had suffered PCP at any point.

On March 20, 1987, without a phase 3 study, and only two years after first being tested, the FDA approved AZT for use. The cost of AZT treatment was eight to ten thousand dollars a year (meaning it could be used only in wealthy Western countries), and there is little doubt that Burroughs Wellcome made millions of dollars profit from the drug.

EQUIPOISE

In stopping the phase 2 trial of AZT early, the researchers had faced a dilemma. It made AZT more quickly available for everyone, but the chance of assessing its long-term effects under controlled conditions had been lost. The state of uncertainty as to which of the two arms in a clinical controlled study is receiving the best treatment is referred to as "equipoise." It is unethical to con-

163

tinue a trial if one treatment is clearly superior. In the case of AZT, the phase 2 study had been "unblinded" early by NIH's Data and Safety Monitoring Board. They concluded that equipoise no longer held: statistical evidence showed a treatment difference between the two arms of the study.

Equipoise sounds fine as an ideal, but it is not clear how easy it is to realize in practice. Are researchers ever in a genuine state of uncertainty? Even at the start of a controlled trial there must be some evidence of a drug's effectiveness or it would not have been tested in the first place. This dilemma famously faced Jonas Salk. He was so certain that his new polio vaccine worked that he opposed conducting a double-blind placebo study. Such a study, in his view, would mean that some people would unnecessarily contract polio. Salk's position was challenged by other researchers who claimed that, in the absence of such a study, the vaccine would not achieve broad credibility among doctors and scientists (Epstein, 201). It is obvious that the idea of equipoise involves complex social and political judgments concerning the credibility of a drug; furthermore, such judgments are made by *researchers*, on behalf of *patients*.

PATIENTS AS BODY COUNTS

Patients in clinical trials are not passive research subjects. In the United States, clinical trials have always been used by patients to get early access to experimental drugs. With AIDS activists making information about new drugs available almost as soon as they left the laboratory bench, patients clamored to take part in AIDS clinical trials.

Two facets of the AZT trials were particularly disturbing to AIDS activists. Because the control group received placebos, this meant that in the long run the only way to tell whether a trial was successful or not was whether the body count in the placebo arm of the study was higher than in the other arm. In blunt terms, a successful study required a sufficient number of patients to die. This they considered unethical. A second criticism was of the rigid protocols of such studies; these forbade participants to take any other medication, even drugs which might prevent deadly opportunistic infections from setting in.

Researchers engaged in such trials were quick to point out that the use of placebos was often the fastest route to learning about the effectiveness of a new drug, and thus, in the long run, saved lives. They cited cases where claims had been made for the benefits of new drugs which randomized control trials had later shown to be useless and sometimes harmful. In response, activists pointed out that there were other options for controlled trials which did not use placebos. For instance, data from treatment groups could be compared with data from matched cohorts of other AIDS patients or patients in studies could be compared against their own medical records. Such procedures were increasingly used in cancer research.

With the growing AIDS underground supplying drugs to patients in desperate circumstances, how accurate was the ideal scenario of the perfectly controlled clinical trial anyway? Although researchers did independent tests to assure compliance (e.g., monitoring patients' blood for illicit medications) and claimed that, in general, patients did follow the protocols, the word from the real world of AIDS trials was somewhat different. Here is how Epstein describes matter:

> Rumors began to trickle in from various quarters: some patients were seeking to lessen their risk of getting the placebo by pooling their pills with other research subjects. In Miami, patients had learned to open up the capsules and taste the contents to distinguish the bitter-tasting AZT from the sweet-tasting placebo. Dr. David Barry, the director of research at Burroughs Wellcome, complaining implausibly that never before in the company's history had any research subject ever opened up a capsule in a placebo-controlled trial,

165

quickly instructed his chemists to make the placebo as bitter as AZT. But patients in both Miami and San Francisco were then reported to be bringing their pills in to local chemists for analysis. (Epstein, 204)

REDEFINING THE DOCTOR-PATIENT RELATIONSHIP

"Noncompliance" is a long-standing concern among medical professionals. But what was happening in the case of AIDS patients was something more radical—the patients, or as they preferred to be called, "people with AIDS"—were renegotiating the doctor-patient relationship into a more equal partnership. The feminist health, self-help movements of the sixties and seventies had already shown what could be achieved. With many gay doctors (some HIV-positive) in the gay communities, such a redefinition could be seen to be in the interest of both doctors and patients.

The new partnership meant that patients had to start to learn the language of biomedicine. Many were already well educated (although not in the sciences), and no doubt this helped. Here is how one AIDS patient described this process: "I took an increasingly hands-on role, pestering all the doctors: No explanation was too technical for me to follow, even if it took a string of phone calls to every connection I had. In school I'd never scored higher than a C in any science, falling headlong into literature, but now that I was locked in the lab I became as obsessed with A's as a premed student. Day by day the hard knowledge and raw data evolved into a language of discourse" (Epstein, 207).

Here is how doctors witnessed the same process: "You'd tell some young guy that you were going to put a drip in his chest and he'd answer: 'No Doc, I don't want a perfusion inserted in my sub-clavian artery,[2] which is the correct term for what you proposed doing'" (quoted in Epstein, 207). As AIDS patients acquired more and more information about the disease, it became increasingly hard in clinical trials to separate out their roles as "patients" or "research subjects" from that of co-researcher.

The criticism by activists of placebo studies returned in 1987 when it became clear that AZT and other antiviral agents might be more effective when given early, well before any symptoms appeared. A number of clinical trials began on early administration of AZT, using placebo groups as controls. Researchers carrying out the trials felt that the criticism of placebos in this case was balanced by the potentially toxic effect of AZT. It was uncertain whether AZT would be effective when given so early, and *not* getting the toxic AZT in the placebo arm of the study might actually be beneficial to the subjects' health. Participants with AIDS, however, saw it differently, especially as they could not take their normal medication during the trial. As one person who discovered he was in the placebo arm of a trial said, "Fuck them. I didn't agree to donate my body to science, if that is what they are doing, just sitting back doing nothing with me waiting until I get PCP or something" (quoted in Epstein, 214). This person also freely admitted that during the study he had taken illicit drugs smuggled in by the AIDS underground. Community doctors were appalled that seriously ill patients were not allowed to take their medicines because of the rigid protocols of such trials. Such doctors became increasingly sympathetic to activists as they tried to find ways out of the dilemma of being both patients and research subjects.

COMMUNITY-BASED TRIALS

Patient groups and community doctors eventually came up with a solution that was as simple as it was radical. They started working together to design their own trials. These would avoid the bureaucratic delays which faced the official trials; they would avoid the ethically dubious practice of using placebos; and, because of the close working relationship between doctors and patients, they had the potential to assure much better compliance. In the mid-1980s two community-based organizations, one in San Francisco and one in New York, despite much official skepticism, started tri-

167

als of new drugs. Such initiatives were particularly suitable for small-scale studies which did not require high-tech medical equipment. The new initiatives also found an unexpected ally in drug companies, who were getting increasingly impatient with the bureaucratic delays in the official NIAID tests.

One of the first successes of community-based trials was the testing of aerosolized pentamidine for the treatment of PCP. NIAID planned to test this drug, but preparatory work had taken over a year. Meanwhile activists had pleaded with Fauci, head of NIAID, to draft federal guidelines approving the drug. Fauci refused, citing lack of data on its effectiveness. The activists came back from the meeting declaring, "We're going to have to test it ourselves." And test it they did. Denied funding by NIAID, the community-based groups in San Francisco and New York tested the drug without using any placebos. In 1989, after carefully examining that data, the FDA approved the use of aerosolized pentamidine. This was the first time in its history that the agency had approved a drug based solely on data from community-based research (Epstein, 218).

There is no doubt that the activists, in an unholy alliance with drug companies and deregulators, had been putting increasing pressure on the FDA. But none of this should take anything away from the activists' scientific achievements, which were significant in terms of how we think about scientific expertise. As a lay group they had not only acquired sufficient expertise in the science of AIDS, but they had, with the help of doctors, actually been able to intervene and conduct their own research. Furthermore, their research had been deemed credible by one of the most powerful scientific-legal institutions in the United States—the FDA.

Part 2

Having traced the early history of activist involvement and documented their first major success, we now describe in part 2 how the activists' own expertise was increasingly enlisted in the science of clinical trials. The radical critique of clinical trials offered by the activists was, as we shall see, eventually to become the established view, signaling a remarkable change in medical opinion as to how such trials should be conducted.

In the mid 1980s a new organization of AIDS activists took center stage. ACT UP (the AIDS Coalition to Unleash Power) was soon to have chapters in New York and San Francisco as well as other big U.S. cities. By the 1990s there were groups in cities in Europe, Canada, and Australia, as ACT UP became the single most influential AIDS activist group.

ACT UP practiced radical street politics: "No business as usual." A typical ACT UP demonstration took place in fall 1988 on the first day of classes at Harvard Medical School. Equipped with hospital gowns, blindfolds, and chains, and spraying fake blood on the sidewalk to the chant of, "We're here to show defiance for what Harvard calls 'good science'!" the activists presented Harvard students with an outline of a mock AIDS 101 class. Topics included:

- PWA's [People With AIDS]—Human Beings or Laboratory Rats?
- AZT—Why does it consume 90 percent of all research when it's highly toxic and is not a cure?
- Harvard-run clinical trials—Are subjects genuine volunteers, or are they coerced?
- Medical Elitism—Is the pursuit of elegant science leading to the destruction of our community? (Epstein, 1)

One of ACT UP's political messages was that AIDS was a form of genocide by neglect. Because of the Reagan government's indifference and intransigence, AIDS was spreading with no cure in sight other than the highly toxic AZT. One of the first targets of ACT UP was the FDA—the "Federal Death Agency," as it was branded by the activists. The culmination of a campaign of protest

169

came in October 1988 when a thousand demonstrators converged on FDA headquarters. Two hundred demonstrators were arrested by police wearing rubber gloves. The subsequent media attention and negotiations with the FDA meant that the government recognized for the first time the seriousness and legitimacy of the activists' case.

Unlike other activist protest movements, such as those in favor of animal rights, ACT UP did not see the scientific establishment as the enemy. In public they applied pressure via well-publicized protests; in private, however, they were quite prepared to engage with the scientists and argue their case. It was the scientists, indeed, to whom the activists increasingly turned their attention. The FDA had been a potent symbol in terms of mobilizing popular opinion, but what really mattered now was getting access to NIAID and the NCI and persuading them to conduct clinical trials in a different manner. It meant engaging with what the FDA called "good science."

TALKING GOOD SCIENCE

AIDS Treatment News heralded the new agenda by declaring in 1988, "The more important question is what treatments do in fact work, and how can the evidence be collected, evaluated, and applied quickly and effectively" (quoted in Epstein, 227).

Over the next three years the activists developed a three-pronged strategy: (1) force the FDA to speed up approval of new drugs; (2) expand access to new drugs outside of clinical trials; and (3) transform clinical trials to make them "more humane, relevant and more capable of generating trustworthy conclusions" (quoted in Epstein, 227). Points (2) and (3), in particular, represented a departure from the standard way of thinking about clinical trials. The normal way of enticing patients to take part in clinical trials was to create conditions under which bribery would be effective: access to drugs was restricted outside of the trials. In contrast,

AIDS activists saw restricted access as the cause of many of the difficulties encountered in clinical trials. As Delaney argued:

> The policy of restriction . . . is itself destroying our ability to conduct clinical research. . . . AIDS study centers throughout the nation tell of widescale concurrent use of other treatments; frequent cheating, even bribery to gain entry to studies; mixing of drugs by patients to share and dilute the risk of being on placebo; and rapid dropping out of patients who learn that they are on placebo. . . . Such practices are a direct result of forcing patients to use clinical trials as the only option for treatment. . . . If patients had other means of obtaining treatment, force-fitting them into clinical studies would be unnecessary. Volunteers that remained would be more likely to act as pure research subjects, entering studies not only out of a desperate effort to save their lives. (Epstein, 228)

This was a clever argument, because rather then reject the idea of clinical trials, it suggested a way in which such trials could be made more reliable. To press this argument required the activists more and more to enter the specialist territory of the scientists and medics—in effect the activists had to tell the medical establishment how to run their trials better.

The activists, as has been pointed out, often started with no scientific background. Remarkably, they quickly acquired a new kind of reputation. They were seen by the physicians and scientists to possess formidable knowledge and expertise about AIDS and its treatment. Practicing physicians were some of the first to encounter the newfound experts. Soon the activists found that physicians were turning to *them* for advice. The director of a New York City buyers' club was quoted as saying "When we first started out, there were maybe three physicians in the metropolitan New York area who would even give us a simple nod of the head. . . . Now, every day, the phone rings ten times, and there's a physician at the other end wanting advice. [From] me! I'm trained as an opera singer" (quoted in Epstein, 229).

Some activists, of course, did have medical, scientific, or pharmacological backgrounds, and such people soon became indispensable as teachers of raw recruits. But most of the leading figures were complete science novices. Mark Harrington, leader of ACT UP New York's Treatment and Data Committee, like many activists, had a background in the humanities. Before entering ACT UP he was a scriptwriter: "The only science background that might have proved relevant was [what I had] when I was growing up: my dad had always subscribed to *Scientific American,* and I had read it, so I didn't feel that sense of intimidation from science that I think a lot of people feel" (quoted in Epstein, 230). Harrington stayed up one night and made a list of all the technical words he needed to understand. This later became a fifty-three-page glossary distributed to all ACT UP members.

Other activists were overwhelmed when they first encountered the technical language of medical science, but they often reported that, like learning any new culture or language, if they kept at it long enough things started to seem familiar. Here is how Brenda Lein, a San Francisco activist, described the first time she went to a local meeting of ACT UP:

> And so I walked in the door and it was completely overwhelming, I mean acronyms flying, I didn't know *what* they were talking about. . . . Hank [Wilson] came in and he handed me a stack about a foot high [about granulocyte macrophage colony-stimulating factor] and said, "Here, read this." And I looked at it and I brought it home and I kept going through it in my room and . . . , I have to say I didn't understand a word. But after reading it about ten times. . . . Oh this is like a subculture thing; you know, it's either surfing or it's medicine and you just have to understand the lingo, but it's not that complicated if you sit through it. So once I started understanding the language, it all became far less intimidating. (quoted in Epstein, 231)

The activists used a wide variety of methods to get enculturated in the science. These included attending scientific conferences, reading research protocols, and learning from sympathetic professionals both inside and outside the movement. The strategy used was often that of learning, as one activist called it, "ass-backward." They would start with one specific research proposal and they would then work back from that to learn about the drug mechanism and any basic science they would need. The activists considered it a sine qua non of their effective participation that they would need to speak the language of the journal article and conference hall. In other words they saw the need to challenge the established experts at their own game. In this, it seems, they were remarkably effective—once researchers got used to their rather unusual appearance. Epstein describes Brenda Lein's experience: "'I mean, I walk in with . . . seven earrings in one ear and a Mohawk and my ratty old jacket on, and people are like, "Oh great, one of those street activists who don't know anything" . . .' But once she opened her mouth and demonstrated that she could contribute to the conversation intelligently, Lein found that researchers were often inclined, however reluctantly, to address her concerns with some seriousness" (Epstein, 232).

Or, as one leading authority on clinical trials commented, "About fifty of them showed up, and took out their watches and dangled them to show that time was ticking away for them . . . I'd swear that the ACT UP group from New York read everything I ever wrote. . . . And quoted whatever served their purpose. It was quite an experience" (Quoted in Epstein, 232).

There is no doubt that some AIDS scientists in their initial encounters were hostile toward the activists. Robert Gallo, the codiscover of HIV, is reported as saying: "I don't care if you call it ACT UP, ACT OUT or ACT DOWN, you definitely don't have a scientific understanding of things" (Quoted in Epstein, 116). Gallo later referred to activist Martin Delaney as, "one of the most impressive persons I've ever met in my life, bar none, in any field. . . . I'm not the only one around here who's said we could use him in the labs." Gallo described the level of scientific knowledge attained by certain treatment activists as "unbelievably high": "It's frightening sometimes how much they know and how smart some of them are" (Quoted in Epstein, 338).

By 1989 the activists were beginning to convince some of the most powerful scientists of their case. No less a person than Anthony Fauci, the head of NIAID, started a dialogue. Fauci told the *Washington Post,* "In the beginning, those people had a blanket disgust with us. . . . And it was mutual. Scientists said all trials should be restricted, rigid and slow. The gay groups said we were killing people with red tape. When the smoke cleared we realized that much of their criticism was absolutely valid" (Quoted in Epstein, 235). The weight of the activist arguments were finally realized at the Fifth International Conference on AIDS held in Montreal in June 1989. Protesters disrupted the opening ceremony, demonstrated against particular profit-hungry pharmaceutical companies, and presented formal posters on their views of drug regulation and clinical trials. Leading activists met with Fauci and enlisted his support for their notion of "Parallel Track." Under this scheme drugs would be made available to patients reluctant to enter a clinical trial while at the same time the trial would go ahead. Scientists worried whether this would mean fewer patients in trials, but patients did continue to enroll in trials even after Parallel Track was adopted.

Activists also sowed doubt about some of the formal rules governing randomized controlled trials. The crucial breakthrough again came at the Montreal conference, where ACT UP New York prepared a special document critical of NIAID trials. Susan Ellenberg, the chief biostatistician for the trials, recalled seeking out this document in Montreal: "I walked down to the courtyard and there was this group of guys, and they were wearing muscle shirts, with earrings and funny hair. I was almost afraid. I was really hesitant even to approach them" (Quoted in Epstein, 247).

Ellenberg, on actually reading the document, found, much to

her surprise, that she agreed with some of the activists' points. Back at her lab she immediately organized a meeting of medical statisticians to discuss the document further. This was apparently an unusual meeting. As she remarked, "I've never been to such a meeting in my life" (Quoted in Epstein, 247).

Another participant said, "I think anybody looking at that meeting through a window who could not hear what we were saying would not have believed that it was a group of statisticians discussing how trials ought to be done. There was enormous excitement and wide divergence of opinion" (Quoted in Epstein, 247).

So impressed were the statisticians by the activists' arguments that members from ACT UP and other community organizations were invited to attend regular meetings of this group. The debate was essentially over whether "pragmatic" clinical trials, which took into account the messy realities of clinical practice, might actually be more desirable scientifically. Here the activists tapped into a long-running controversy among biostatisticians as to whether clinical trials should be "fastidious" or "pragmatic." A pragmatic clinical trial works under the assumption that the trial should try to mirror real-world untidiness and the heterogeneity of ordinary clinical practice patients. Such pragmatic considerations were already familiar to some biostatisticians who had experience of cancer trials, where different and more flexible ways of thinking about trial methodology had already been instituted. The fastidious approach favored "clean" arrangements, using homogeneous groups. The problem with the fastidious approach was that although it might deliver a cleaner verdict, that verdict might not apply to the real world of medical practice, where patients might be taking a combination of drugs.

THE EXPERTNESS OF LAY EXPERTISE

What particular expertise did the activists possess? Or did they just have political muscle? Scientists are predisposed to avoid 175 political interference in their work, especially by untrained outsiders. Without any expertise to offer, the politicking of the activists would, if anything, have made the scientists more unsympathetic.

The activists were effective because they had some genuine expertise to offer and they made that expertise tell. In the first place, their long experience with the needs of people with AIDS meant that they were familiar with the reasons subjects entered studies and how they could best be persuaded to comply with protocols. Fauci described this as "an extraordinary instinct . . . about what would work in the community . . . probably a better feel for what a workable trial was than the investigators [had]" (Quoted in Epstein, 249). Activists also had a particularly valuable role to play as intermediaries, explaining to people with HIV and AIDS the pros and cons of particular trials.

But the expertise went beyond this. By learning the language of science the activists were able to translate their experience into a potent criticism of the standard methodology of clinical trials. By framing their criticisms in a way that scientists could understand they forced them to respond. This was something which the Cumbrian sheep farmers discussed in *Golem at Large* were unable to do. The activists were lucky because at the very time they were raising these concerns, some biostatisticians were reaching broadly similar conclusions themselves.

One of the most fascinating aspects of the encounter between the activists and the scientists was the give-and-take on both sides. For example, as the activists learned more and more about the details of clinical trials, they started to see why, in some circumstances, a placebo study might be valuable. Thus, in a panel discussion in 1991, AIDS activist Jim Eigo acknowledged that although originally he had seen no need for placebos, he now recognized the virtues of using them in certain situations where a short trial could rapidly answer an important question.

While AIDS activists embraced the "real-world messiness" model of clinical controlled trials, some were chastened by the experience of carrying out real-world research. Martin Delaney of Project Inform admitted, after conducting a controversial clinical trial without any placebos, that "the truth is, it does take a lot longer to come up with answers than I thought before" (Quoted in Epstein, 258).

TEACHING OLD DOGS NEW TRICKS

The success of the activists' arguments about the need to make clinical trials more patient-oriented was marked in October 1990 by the publication of two back-to-back "Sounding Board" articles in the *New England Journal of Medicine*. One, by a group of prominent biostatisticians, argued for restructuring the phases of the FDA approval process, dismissed the requirement of homogeneity in clinical trial populations, and called for more flexible entry criteria. It concluded by calling for patients to participate in the planning of clinical trials. The second article, by a well-known Stanford AIDS researcher, was titled "You *Can* Teach an Old Dog New Tricks: How AIDS Trials are Pioneering New Strategies" and took up similar themes of flexibility, and how to ensure that each limb of a trial offered benefits to patients. Medical ethicists soon came on board supporting the new consensus about how AIDS trials should be conducted. And indeed, trials started to be conducted according to the protocols suggested originally by the activists. Another victory came when NIAID began recruiting an increasingly diverse population into trials.

By the next international AIDS conference the activists were so well accepted into the AIDS establishment that they spoke from the podium rather than shouting from the back of the room (Epstein, 286). In a speech at the conference Anthony Fauci announced, "When it comes to clinical trials, some of them are better informed than many scientists can imagine" (Quoted in Epstein, 286).

The success of the activists in speaking the language of science had one paradoxical outcome: it meant that new generations of activists increasingly felt alienated from the older activists. Indeed, splits and tensions appeared between "expert lay" activists and "lay lay" activists. As one New York activist reflected, "There were people at all different points within the learning curve. . . . You'd have somebody . . . who had AIDS, who knew a lot about AIDS, [but who], didn't know *anything* about AIDS research—you know, nothing. And never had seen a clinical trial, didn't live in a city where they did clinical trials, on the one end—and then Mark Harrington and Martin Delaney on the other" (Quoted in Epstein, 293).

This division of expertise is exactly what we would expect given the Golem model of science. Expertise is something that is hard-won in practice. Someone with AIDS might be an expert on the disease as it affects patients, but this does not make him an expert on the conduct of clinical trials. It was the particular expertise of the activists which commanded the scientists' attention, and if new activists were going to have any influence, they too would need to become experts.

Some AIDS activists, as they became increasingly enmeshed in the science of AIDS, even became "more scientific than thou" as we might say, when it came to assessing treatments. On one famous occasion, well-known activists chastised a leading AIDS researcher for trying to make the best of clinical trial data by post hoc redrawing of the sample into subgroups in order to claim some effect. Some activists refused to countenance alternative therapies for AIDS, and some even enrolled in medical school in order to undertake formal scientific training.

The activists were not a homogeneous body, and splits and tensions arose between different groups and within groups. The New York activists in comparison to the San Francisco group were seen as being more closely aligned to orthodox science. But even the

New York activists always maintained that a core part of their expertise lay in their being part of the community who were living with and dying from AIDS. It was their experiences of the world of the patients which gave them something medical experts could not hope to acquire unless they themselves had AIDS or were part of the gay community.

Although activists went on through the 1990s to contribute to other issues, such as debates over combination therapies and the role of surrogate markers in assessing the severity of AIDS, it is in the arena of clinical trials that their most stunning victories were achieved. In effect a group of lay people had managed to reframe the scientific conduct of clinical research; they changed the way it was conceived and practiced.

This success shows us that science is not something that only qualified scientists can do. Just as lay people can gain expertise in plumbing, carpentry, the law, and real estate, they can gain expertise in at least some areas of science and technology. In some areas, they may already have more relevant experience than the qualified experts. But the crucial issue, as we have seen in this chapter, is getting that expertise recognized as such. And it is this more than anything which the AIDS activists have been able to achieve.

Vaccination and Parents'
Rights *Measles, Mumps,*
Rubella (MMR), and Pertussis

Nowhere is the tension between individual choice and the collective good more marked than in the case of the vaccination of children. Fairly complete vaccination of populations can eradicate some diseases entirely. Thus has the terrible threat of smallpox been removed from the planet. So thoroughly has it been eliminated that barring accident, terrorist attack on the one or two places it is preserved in vitro, or its spontaneous reemergence or reinvention, no one will ever again need to be vaccinated against it. But to eradicate a disease only a high proportion of the population—not 100 percent—needs to be vaccinated. So if you as an individual do not like the idea of vaccination, you could just wait until enough others have been vaccinated to bring about the "herd immunity" that will protect you. Unfortunately, the more people who take that view, the less likely is the disease to be eradicated—the less likely is herd immunity to be achieved—because the more potential carriers remain in the population.

This is a classic situation whose logic is like that of the famous Prisoner's Dilemma: you never quite know whether you will maximize your rewards by going straight for what you want or by asking for a little less—it all depends on what others do. Imagine two prisoners who cannot communicate with one another. Each is told, "If you betray the other person and he/she doesn't betray you, you'll go free, and they'll get ten years. If you betray them and they betray you, you'll both get ten years in prison. If neither of you betrays the other one, you'll each get one year in prison." Think of vaccination as equivalent to a year in prison and catching the disease in question as equivalent to ten years. If everyone vaccinates, then everyone gets one year. If no one vaccinates, then everyone gets ten years. If everyone else vaccinates and you do not, you go free.

MMR

The equivalent of the one year in prison in the case of the measles, mumps, and rubella vaccine is actually something much worse. Any parent would accept a year in prison in exchange for having their child avoid autism—the claimed consequence of the MMR injection in a subset of cases. If you are a parent, how can you volunteer your child for a jab that might have such a devastating effect on your child's life?

But is there something still worse than exposing your child to the risk of autism? Is there an equivalent to the ten years in prison? The answer is that there is—the equivalent is measles. In a small number of cases measles causes severe brain damage. Whether the brain damage caused by measles in any one case will be worse than autism is hard to say; among other things it would depend on the severity in each case. But the likelihood of your child getting brain damage should they catch measles is greater than the likelihood of your child becoming autistic as a result of having the MMR vaccine.

Of course, the ideal situation is to avoid both. You can accomplish this if you refuse the MMR vaccine while enough others take **181** it to suppress the epidemic. That is the equivalent of no time in prison at all. The trouble is that if too many parents think like that, and there is a measles epidemic, then many children will get measles and brain damage as a result of your actions, and your own child's chance of getting measles becomes high.

But this is to make the calculation simple by overlooking a deeper problem. The deeper problem is whether there is any risk at all of autism following MMR vaccination. This is what we know: a number of children start to show the symptoms of autism in the first few years of life, around the time that the MMR vaccine is typically administered. Given that the onset of autism and the administration of MMR happen in the same period of the child's life, autism onset will sometimes precede the administration of the vaccine and sometimes follow it. Now, consider those vaccinated after the onset of autism; because of what everyone already knows about the way the world works, nobody is going to think the onset of autism caused the subsequent MMR vaccination. On the other hand, in those cases where the autism followed the vaccination it is quite reasonable to think that the vaccination caused the autism. And parents are still more likely to see a purely temporal sequence as a causal sequence if the idea that there might be a causal link becomes widely broadcast. Hence, whether or not there is a causal link, it is easy for parents to become convinced that autism in their children was caused by the vaccination that preceded it.

In the case of the recent panic over MMR, the parents' concerns were boosted by journalists who, looking for "balance" in their stories, tend to present the issue as one of medical experts against parents, giving the parents' views equal weight to that of the experts. Thus, the *Western Mail* (which advertises itself as being "The National Newspaper of Wales"), carried the following headline to the front-page lead story on September 5, 2002: MUM CLAIMS NEW MMR AUTISM LINK. The first few paragraphs of the story read as follows:

Fresh evidence emerged last night to suggest that the MMR vaccine is linked to autism.

The parents of Welsh schoolboy [name] have discovered that his blood and digestive organs are infected with the same strain of measles used in the triple vaccine. And they fear his condition will get worse if the disease has spread to his brain.

[Name, age,] was diagnosed with autism and a severe bowel disorder when he was two, soon after having the MMR jab to protect him against measles, mumps, and rubella. Last night his mother [name], who lives near Newport, said she believed there was no other way he could have been infected by measles except through the jab he was given as a toddler.

"We more or less knew this was the case because to my knowledge [name] has never been exposed to this strain of measles except through the vaccine," said Mrs [name].

The discovery of the virus consistent with a strain of measles used in the MMR vaccine was made after specialist tests. Now [name's] condition is certain to cause concern among other parents being asked to give their children the triple jab.

Mrs [name] has always maintained that her son's illness was caused by an adverse reaction to the MMR vaccine.

It turns out that the initial medical evidence for the existence of a causal link was developed as a result of parental worries of this kind presented before any scientific research on this matter had been reported. In the first paper on the subject the authors (of whom the principal author is Andrew Wakefield) thank "the parents for providing the initial impetus for these studies."[1] This paper reports twelve cases of children with behavioral abnormalities which developed after the MMR jab, the link being noticed by parent or doctor in eight of those cases. In the eight cases the gap between vaccination and onset of symptoms was reported as, respectively, one week, two weeks, forty-eight hours, "immediately after," one week, twenty-four hours, two weeks, and one week. It is easy to see that distraught parents, seeking to understand why they and their child had suffered this tragedy, would seek out the most

183

salient event in the child's life in the immediately preceding period. But the authors of the paper concede that they "did not prove an association between measles, mumps, and rubella vaccine and the syndrome described," as indeed they could not with evidence of this kind. The majority of the paper is concerned with establishing a link between abnormalities of the bowel and intestines and the behavioral disorder. Whether this association is reasonably established is not known to us, but even if it is, the relationship between the vaccination and the intestinal disorders is not.

In the absence of some reasonably established causal chain from MMR vaccination to intestinal disorder to behavioral disorder—a causal chain that is unlikely to be established for many years—one needs other kinds of evidence. The other kind of evidence can be found in the statistics of the whole population rather than the features of any one case or a small number of cases. In other words, once one understands the background numbers—the spread of age of onset and distribution of autism in the population, and the pattern of injections with MMR vaccine—one can ask if there are significantly more cases of the onset of autism after vaccination than before it. Furthermore, we can look at the changes in patterns of autism consequent on the introduction of the vaccine to whole populations.

The original paper admits that there is evidence that autism has a genetic origin—more boys have it than girls, and both of a pair of monozygotic twins are more likely to show the symptom than both of a pair of dizygotic twins. When we move on to population statistics, the author, Andrew Wakefield, says that the evidence is unclear, though most others seem to think that it goes against the link.

Epidemiologists, those who study the incidence of diseases in whole populations, find that there is no measurable excess of the incidence of autism associated with MMR. This suggests very strongly that the parents are wrongly imputing cause from sequence. Of course, an epidemiological study, in virtue of its statistical nature, cannot eliminate the possibility that a very, very, small number of children have been affected by the vaccine. It may be that one or two parents are right to infer that the jab was the cause of the autism condition in their children. But even if their concerns do have some basis, it is almost certain that the risk to the children of exposure to a disease such as measles is much greater than the risk associated with the vaccine. The risk from the vaccine is too small to be visible in the population statistics, whereas the risk associated with measles stands out.

In spite of the statistics, the concerns of parents, given salience and concreteness by the press, are reinforced by Web sites that campaign against vaccination in general and MMR in particular. A survey of antivaccination Web sites in June 2002 found a total of twenty-two such sites.[2] The authors' conclusion was that these sites "express a range of concerns related to vaccine safety and varying levels of distrust in medicine. The sites rely heavily on emotional appeal to convey their message." They found that 55 percent of the sites contained "emotionally charged stories of children who had allegedly been killed or harmed by vaccines" (3245). A quarter of the sites included pictures of distressed children. As the authors remark, "Such visual images of purported adverse consequences can be unsettling to parents facing vaccination decisions" (3247). They point out that nowadays there may be a prevalence of such images because "the once overwhelmingly apparent visual images of the benefits from vaccination have disappeared as their respective diseases—such as polio—have disappeared"(3247). A number of these sites were also concerned with the "big brother" aspect of enforced vaccination and the morality of using aborted fetuses as a source of vaccine material.

By 2002 Andrew Wakefield, the principal author of the original paper associated with the anti-MMR movement, had been pressing forward the case against the vaccine for five years or so but had

185

failed to build a significant body of support from among the medical profession. That is to say, Wakefield could find supporters for his claim that measles virus in the gut was associated with autism, but not for his inference, made at a news conference, that MMR per se could be a cause of autism. Wakefield, it should be remembered, knowing the epidemiological evidence, was still recommending that parents vaccinate their children against measles, so it is only the particular association with MMR that is at stake. It could be said, then, that there was no real dispute among experts in this case, understood as the MMR per se case; the entire debate was between the medical profession and the public, bolstered by journalists and the Web.

Discussion

The MMR debate is an almost perfect example of the problems of the interaction between scientific medicine and the public. One day we may know exactly how vaccines interact with the body. As we argued in the chapter on the placebo effect, one day we may understand the vaccines at the level of the cell in the same way that today we understand broken bones at the level of the individual bone. We may discover that there is indeed a very small set of people who experience extreme reactions to vaccines because of their genetic inheritance in a way that the general population does not. The epidemiological evidence we already have tells us that this is bound to be a very small number, but once we can identify individual members of the subset everything will change. To administer a vaccine to a member of that very small group would be like feeding nuts to someone with a deadly allergy; it would never be considered.[3]

But we will not be in a position to understand the threat of vaccination on an individual-by-individual basis for many years to come. All we have now is statistical medicine that works on populations, the same medicine that forces us to use double-blind control trials and placebos. Right now we have no choice but to make decisions based on what we know. This is not an untypical situation in science. As we argue in the other volumes of the *Golem*, scientific and technological disputes often take many decades to settle, while the pace of decision-making about matters of public good is, perforce, much quicker.

Sometimes the solution to technological decision-making under conditions of uncertainty is to resort to the "precautionary principle." The precautionary principle says that if the risks of a certain technological innovation are not yet fully understood, the wise course of action is caution—do nothing. This is a powerful argument in the case of, say, genetically modified foods and other organisms (GMOs). It may not be a decisive argument, because the benefits of genetically modified crops to third-world farmers may be so huge that simply doing nothing may be causing people in the third world to die of starvation, but it is a strong argument. Here, however, the precautionary principle does not help us. We know for sure, or almost for sure, that if we stop vaccinating there will be epidemics and that the consequences of the epidemics will be worse than the consequences of the vaccination.[4]

We have, instead, to concentrate our minds on the choices that parents face, and we cannot wait until the science of vaccination reaches a level equivalent to the science of broken bones. Here the uncertainties of medicine are immediate problems, just as we anticipated in the preface. Somebody is making a decision right now about whether to vaccinate his or her child.

Here the agencies responsible for the collective can act in only one way. They must encourage or enforce vaccination for dangerous diseases at least insofar as this is compatible with the political realities. There is no scientific evidence that would absolve them from the charge of irresponsibility were they to act in any other way. Those agencies cannot be certain that there is no link between MMR and autism in a very small number of cases, only that

187

all the established evidence—the pieces of evidence that are candidates to be counted as secure scientific findings—point the other way. This is not to say that this holds true for vaccinations for all diseases or for the tendency to vaccinate for more and more diseases, but there seems little room for governments to act in any other way in the case of MMR.[5]

Or perhaps there is. On February 6, 2002, at around 8.30 A.M., spokesmen for the British governing Labour Party minister of health and the Conservative Party shadow minister debated the issue on the BBC *Today* program.[6] The Labour Party spokesman took the line recommended here. He made clear the party's view that there was no evidence for any danger associated with MMR in either Britain or any other country and said that they would go ahead with the existing vaccination policy. The shadow minister took another view. Agreeing with the Labour minister on the evidence, he nevertheless wanted to grant parents freedom of choice. One of the alternatives to MMR demanded by some parents was the separate administration of the three vaccines. The Labour Party was resisting this on the grounds that there was no evidence that there was any need for a change of policy and, presumably, that separate administration was less effective on the grounds of cost, demands on doctors' and parents' time, the reduced likelihood of complete take-up of one or more of the vaccines, as it is easier to forget to administer three vaccines than one, and the fact that the chances of an epidemic would be increased because children would, on average, be vaccinated against these diseases later and thus be exposed for longer.[7]

The Conservative spokesman made clear that his party still believed that MMR was safe, but that take-up of MMR was now so low that measles epidemics were breaking out. He therefore argued that parents should be offered the choice of the single measles vaccine in an attempt to increase take-up and reduce the chance of an epidemic.

Here we see that in some ways the MMR debate cannot be treated in isolation. In Britain the government has a poor track record in respect of science. Governments (and it tends to be Conservative governments simply because they were in power for seventeen years prior to 1997), have made a series of mistakes. The spectacle of John Gummer, the spokesperson for the Conservative government responsible for handling the BSE (bovine spongiform encephalopathy) outbreak, feeding a hamburger to his young daughter in front of the television cameras to prove that British beef was safe, looms large in the recollections of every British subject more than about twenty years old. The Conservatives were wrong about this; BSE, it turned out, could cross the species barrier to humans in the form of new variant Creutzfeldt-Jakob disease (CJD), and has so far killed and will continue to kill people who were big eaters of the wrong kind of beef.[8] And, of course, no one with any sense trusts a word that governments say about the safety of nuclear-fueled power plants and the associated industries—governments have been wrong about this too. Thus, the Labour Party's policy—trying to convince the British electorate that their view on the safety of MMR should be accepted—has been made much harder by previous debacles (including their own mishandling of the foot-and-mouth disease outbreak and their unpopular position on genetically modified foods). The spin-offs from these previous episodes make the Conservative policy more reasonable given the political environment, even if it was disreputable in terms of science. What has been established was that the British electorate likes a choice where these large-scale technological decisions are concerned.[9] Who is right in this case? We merely present the arguments, but Collins thinks that presenting the arguments in this way favors the Labour position, because in spite of past failures it is dangerous to legitimate the role of scientifically ill-informed public opinion in debates of this kind, where there is no significant scientific support for it. This is especially the case where, as in the case of vaccinations, the health of the population at large is at issue. Pinch, on the other hand, believes that we as analysts are not in a position to make such a judgment, and that the outcome should be left to the ebb and flow of public opinion within the normal political process.[10]

What about the individual parents? If given a choice, how should they exercise it? Here the logic is, on the face of it, different. Presumably a truly ruthless parent would encourage every one of his friends and acquaintances to believe that vaccination was the only acceptable policy while surreptitiously refusing vaccination for his own children. If other parents were sufficiently gullible, this would avoid the usual penalty of the Prisoner's Dilemma because there would be no epidemic. But how could one possibly recommend such a vilely selfish policy?

In any case, in the long term the selfish policy will fail. Those who study the Prisoner's Dilemma know that the logic unfolds differently if the situation recurs over and over again and we have the "*repeated* Prisoner's Dilemma." That is to say, over time, if the dilemma is repeated, each prisoner learns how the other acts and if the other prisoner always acts selfishly, then selfishness becomes the norm, to the long-term disadvantage of everyone— everyone betrays everyone; everyone gets ten years; epidemics spread.

Still, that is theory and that is the long term. It would be so much easier to act unselfishly if one could believe the scientific evidence that suggests that the impression that autism is caused by MMR vaccine is just that—an impression. Here, then, we have to begin to ask ourselves the difficult questions. What kind of expertise can parents bring to their decision? We can answer this question only in terms of scientific expertise. There are going to be some parents who think the answer will be given to them by prayer or religious teaching, by consulting astrologers or oracles, by consulting practitioners of alternative medical regimes that have no basis in Western science and have undergone no controlled testing of the kind to which Western medicine must aspire. We can offer an answer only for those who consider that they live in a society whose values are broadly aligned with Western science. (This does not involve thoughtlessly embracing all the offerings of Western science, but accepting that when these offerings are rejected it should be on scientific grounds related to the particular case in question, or on the precautionary principle, but not on a blanket rejection of science per se.) We need to ask, then, about what kinds of expertise can be acquired by parents in their quest for a guide to their decision-making. And we will assume that at some point the evidence they need is about the likelihood that MMR will cause behavioral problems in their children.

The Authors of This Book and Whooping Cough (Pertussis)

As it happens, we have some direct evidence about how parents go about judging the risks of vaccination for their children, because both of the authors of this book have done it, not in respect of MMR, but in the case of whooping cough, otherwise known as pertussis. Collins can be dealt with quickly: his children were vaccinated, in Britain, in the late 1970s and early 1980s. At the time there was much discussion among the group of mothers and fathers about the dangers of the vaccine, but most concluded that the dangers of the illness were worse and had their children vaccinated. This is what the Collinses decided and all turned out well for their children. The Pinches, however, represent a much more complicated case. In November 2002, Collins interviewed the Pinches (quite aggressively) about their choice, and Collins subsequently wrote up the account. Here Collins will use the first-person singular and describe the Pinches as subjects even though Trevor Pinch also wears another hat as coauthor of this book.

The Pinches' Decision as a Case Study

In 1992, the Pinches, living in America, had to decide whether to allow their daughter to be given a series of DTP (diphtheria, tetanus, pertussis) vaccinations in the first few months of her life. They decided against it, eventually, settling on a policy of a combination of only two of these vaccines (DT) after one year and finding a way of having a special form of inactive pertussis (said to have fewer side effects) given in combination with DT at one-and-a-half years and repeated at two years. It is the cause and manner of their decision that are the next topic. We are going through this account so that we can use it to reflect back on the still open-ended issue of MMR.

The Pinch parents are both sociologists and both have an interest in the sociology of knowledge. Both work at Cornell University. Trevor Pinch is British. Because he is a sociologist of knowledge, he is not likely to be impressed by arguments that emerge purely from scientific authority; this is not to say that he is in any way "anti-science," but he knows that there can be more uncertainty about scientific claims, and medical claims in particular, than some doctors and scientists will admit. Also, having experience of the German and British health care systems, where vaccination is ultimately a matter of parental choice, Pinch was made suspicious by the U.S. policy of denying unvaccinated children entry to school. The Pinches are also familiar with the medical regime of German-speaking Switzerland, where alternative medical regimes, such as homeopathy, have far more currency and legitimacy than they do in Britain or America. In Ithaca (the remote rural small town which is home to Cornell University), the Pinches found that there were both orthodox and alternative support groups for pregnant women and visited them prior to the birth of their first child. It was the alternative group, a group that continued to meet weekly even after babies were born, that reinforced their doubts about the pertussis element in DTP.

When I asked the Pinches what reasons they could provide for what was, on the face of it, the self-centered choice to delay vaccinating their child, they went to their study and brought out a cardboard box full of old leaflets and papers dating from the early and mid-1990s which they had studied at the time. Many of these leaflets (many of which were supplied by their pediatricians), carried hand-written annotations, the statistics of the chance of side effects of the vaccine and the chance of evil consequences from catching the illness being the main focus of concern.

Trevor Pinch stressed that he himself was skeptical about the antivaccination literature and understood that the pictures of sick children they contained were emotional and misleading. Nevertheless, the standard leaflets, available from the doctor, when read carefully, were a cause for concern.[11] The vaccine that causes most side effects was DTP, which was one the first to be administered in a baby's life. Reading carefully, one could see that it was the pertussis element in the triple vaccine that was the problem. So, in the first instance, says Pinch, "we asked for DT only. And this was administered. They thought this was unusual and they gave us endless arguments about it but eventually they did it. (We said that if we didn't get the DT we would resist all vaccinations.)"

I pressed the Pinches further—after all, it was well known that whooping cough was a dangerous disease. Trevor explained by describing the possible side effects of the vaccination:

> You looked up the statistics and it turns out that most children will have slight fever and irritation, and half of the children get soreness and swelling in the area of the vaccination. In one case in 330, temperatures of 105 or higher are recorded. Continuous crying lasting 3 or more hours after occurs in one case in every 100; high-pitched crying in one in every 900; convulsions, limpness or paleness was one in every 1,750; and the last was for me the crucial statistic.
>
> In the same leaflet it said that for babies that caught pertussis, "as many as 16 out of 100 babies with pertussis get pneumonia and as many as 2 out of 100 may have convulsions." One out of 200 has brain problems and one out of 200 dies.

I said to the doctor, "I look at it like this: It is unlikely that my child will get whooping cough because it is a rare disease." It depends on where you live and your lifestyle. Our child was not being exposed to many other children. Even if she got it, it is only 1 in 200 who get convulsions, but I read in the same leaflet that 2 out of 100 who get the injection get convulsions.[12] I did look up the statistics and found that the chances of my child getting whooping cough were very, very small. The same leaflet said that in recent years 4,200 cases of whooping cough were reported yearly in the USA. The leaflet said "as many as 4,200 cases" but it seemed to us 4,200 (and this included adult cases) was not many in the U.S.—it was a rare disease. (Another leaflet gave the figure of whooping cough cases in the U.S. as only one in 2,000 and only 9 deaths per year.)

It seemed clear that my child was far more likely to get convulsions as a result of being vaccinated than as a result of getting whooping cough. We also found out that whooping cough is not very severe in children over seven years of age, so all we had to worry about was the chance of our child catching it before the age of seven, and this chance was slight.

At this point I challenged the Pinches' reasoning. Surely, I argued, the crucial figure is not the incidence of convulsions; the Pinches had no evidence that convulsions led to long-term adverse health consequences (I could not prove they did not, of course). Surely, I argued, the crucial figure in the leaflets is that if your child catches pertussis he or she has a 1 in 200 chance of permanent brain damage and a 1 in 200 chance of dying, adding up to a 1 in 100 chance of permanent or terminal injury. Though the leaflets admit that death or permanent injury can be a side effect of the vaccine, the chances are so slight that the statistics are not even recorded. "Weren't you irresponsibly exposing your child to a far greater risk by refusing the vaccination?" I asked.

The Pinches' response was that the chances of their child catching whooping cough in rural Ithaca was tiny given the low incidence of the disease in the United States, the tendency for it to be transmitted in crowded and poverty-stricken areas, and the health and strength of their child. They also argued that in the first year of life their child would rarely be taken into public places where she would be exposed. Furthermore, they argued, their child's high birth weight, history of breast feeding, and generally good nourishment made it less likely that even if she caught the illness she would suffer the most extreme side effects.

However valid this line was in terms of the likelihoods, it seemed to me that the Pinches could still be accused of putting the health of their child above the health of children in the community. The existence of more potential transmitters of whooping cough in the population, strong and healthy though they might be, exposed poor and unhealthy children to more potential risk. But it is easy to argue this when it is not your child who is about to be exposed to danger, and, in any case, I suspect that the Pinches would have allowed live pertussis vaccine to be administered had it not been that they knew that another strain of vaccine was available. As Trevor explained:

> We were not arguing that our child should not be vaccinated against whooping cough, but we wanted her to be vaccinated with the inactive version of the vaccine known as the DaPT, which was used in Europe and, we knew, was routinely used in Japan. This we believed was as effective and had far fewer side effects. We must have had a leaflet about that (this was before the days of the Internet). The people in the alternative birth group shared information about this. We were getting piles of information every week from someone in the birth group. It was a collective endeavor.
>
> We did not want to resist all vaccines; we just wanted to go slower, to avoid cocktails of vaccines, and to avoid vaccines with side effects when there were available alternatives that had fewer side effects.

One thing that should be borne in mind was that the Pinches' decision was taken in the context of the campaign to recognize Gulf War syndrome (See chapter 5), which was said by some activists to have been caused by the cocktail of vaccinations given to troops in operation Desert Storm.

Eventually, the Pinches succeeded in accomplishing the vaccination regime they wanted. "We insisted on the DaPT vaccine and eventually they ordered it in especially for us. Our relationship with our doctor was never good, however—she called us 'irresponsible parents.' The nurses and we also had a very strained relationship. It made for a very unpleasant situation because the doctors and nurses are not used to being confronted with their own statistics and they tend to fall back on authority and call you irresponsible; they fall back on the party line."

The Pinches feel vindicated in their choice because "one year after these arguments we read in the *New York Times* that America had decided to start administering the DaPT, as it was the safest vaccine with fewest side effects."

Interestingly, the Pinches also stressed that an important element in their decision was their distrust of the commercial pressures experienced by doctors. For example, they knew that doctors were paid per vaccination, and they found that advisory leaflets were provided by drug companies. They also noted that free medical visits were always scheduled by the pediatricians to occur when a vaccination was due. Collins suggested that the last point bordered on paranoia, since a benevolent medical regime working in a situation of scarcity would schedule visits in this way to maximize take-up of vaccination irrespective of commercial pressures. The Pinches were also worried by the trend in the United States for more and more vaccinations against diseases such as chicken pox which cause only inconvenience rather than danger. The pressure under which parents were placed to complete the mandated vaccinations, including exclusion from school, worked against open discussion and decision-making. Collins, on the other hand, argued that irrespective of the emotive power of these wider political considerations, the decision to vaccinate or not to vaccinate should always be made on the basis of the technical evidence. That is not to say that relations between doctors and parents could not be handled better.

Conclusions on the Pinches Decision

It seems that in this argument the Pinches hold the trump card. Somehow they had managed to anticipate a change in policy by the American government in the direction of their ideas about medicine. Who can argue with that?[13] The Collinses, bear in mind, had no such option, since the DaPT vaccine did not come into use anywhere until long after their children were vaccinated (though this is not to claim that the Collinses would have been so assiduous in their research, or so fortunate in their network of supporters, as the Pinches, even if some other vaccine had been used in another country).

But the Pinches' trump card is not much help when we are considering a case such as MMR where there is no other vaccine being used elsewhere and no evidence that other vaccination regimes, such as one vaccine at a time, are safer. The question we really need to ask is what the Pinches would have done had they not known of the existence of DaPT vaccine—effectively the choice faced by ill-educated parents who had not traveled widely in Europe. The Pinches say that under these circumstances they would have given live vaccine.

As it was, their child's vaccination was delayed by only one year. That being the case, Trevor argues that the chance of permanent or terminal injury was small when compared to the slight risk of serious side effects caused by the vaccine because of the low incidence of whooping cough and the good health of his child; Collins argues that this is essentially an argument that puts the health of the individual child above community considerations. Pinch argues for more choice; Collins argues that given the understandable tension between individual and community interests, an enforcement regime is appropriate. The Pinches argue that their actions were an exemplary case of an educated group fighting for the right decision for their child, vindicated by the subsequent

197

change of policy. The Pinches argue that they knew more about pertussis vaccination than the doctors they encountered. Collins argues that it is a hard matter to judge whether their child and others were exposed to needless risk as a result of the Pinches' judgment.[14]

Now, it happens that ten years later, in 2002, there was an epidemic of whooping cough in Ithaca. This appeared to be the kind of epidemic that breaks out regularly on about a three-year cycle in one location or another. The number of cases in Tompkins County, the area in which Ithaca lies, had increased to just over seventy by November 4, 2002 compared to a typical figure of one or two per complete calendar year.[15] Of the seventy cases, two babies had experienced severe breathing difficulties, though fortunately no permanent injury had resulted. (In the ten years that our informant administrator had worked in Tompkins County no serious side effects had been experience by any child who had been vaccinated.) Thus, the argument against vaccination from the small number of cases in the United States and the rural nature of some environments does not always hold.

The Tompkins County outbreak does not seem to be the result of poor take-up, but provides a graphic illustration of how views could change if epidemics were to become widespread. A premise of our argument is that epidemics develop more easily in unvaccinated populations. For someone who is determined to believe that the decline of infectious diseases has nothing to do with vaccination and is a result of improving nutrition or the normal ecology of disease (in spite of the evidence of the effectiveness of vaccine in the less developed world), then nothing follows. But, if one of the premises of the argument in this chapter is accepted (and it is accepted by almost everyone), namely that vaccination decreases the chance of epidemics, families who took a decision to avoid vaccination for their own children on purely individualistic grounds would find that their grandchildren had a greatly increased chance of catching the disease (with its 1 in 100 probability of permanent or terminal injury), partly as a consequence of their earlier decision to try to protect their offspring. The same would apply to their young nephews, nieces, cousins, and so forth. Thus, even if safeguarding one's own offspring irrespective of the effect on the population as a whole were the object, it would not succeed. In the question of vaccination, in the long term there are no free riders.

How to Make the Vaccination Decision

Here, then, we see an almost perfect example of decision-making by those without medical qualifications. The Pinches were both teachers at one of America's leading universities with training in a subject which bore a relationship, if only a distant relationship, to the science of medicine. And they had done research to the extent of filling a cardboard box with relevant literature, consulted with parents who had collected similar literature, and built up a collective body of knowledge on how to evaluate the literature.

What do these skills comprise? Note, first, what they do not comprise. Neither of the Pinches could get a post in medicine. In our terms, they have no contributory expertise in medicine.

What they do have, however, is what we have been calling interactional expertise, that is, the ability to read and understand the medical literature, reinforced by their social networking with others who also know the area in question, more or less, to the extent that they know how to question doctors and nurses and put alternative points of view to them without appearing simply ill-informed or stupid. In other words, they are capable of putting arguments that skilled medical personnel ought to feel obliged to answer; the arguments cannot (or should not), be dismissed on the basis of authority.

In spite of this almost optimum position, it is not clear that the Pinches made the right decision in avoiding vaccination for their child's first year of life. Their strategy of avoiding early vaccination with DTP and avoiding cocktails of vaccines drew on their knowledge of the side effects (though on Collins's account, an improper balancing of these against the much more severe dangers of the disease itself), and the common-sense idea that the less the assault on the very young child's immune system the better. Here the counterargument is that the new immune system is constantly bombarded with thousands of challenges from the environment and that the vaccines are but a small addition. The Pinches also admit to a general suspicion of medical assaults of that magnitude, fearing that in the state of uncertainty it might even be a cause of serious imbalances in later life (to which Collins counters that the same argument could be used against any traumatic medical intervention, such as swamping a wound with iodine or any other procedure associated with the relatively new germ theory of disease). Collins argues that the cost of avoiding early vaccination and cocktails (and avoiding cocktails inevitably delays the date of vaccination), means that their child was exposed to the risk of catching the illness for longer than other children, a potential problem for their own child, for the community and, in the fullness of time, for their children's children. Spreading the idea of later and longer, drawn-out vaccination programs also increases the chance of low take-up and the chance of other children contracting the disease with the same deleterious consequences for the community and future generations. These points may not be decisive, but they are arguable. Even after all the trouble the Pinches went to, if there had been no DaPT, the Pinches may have been right, or they may have been wrong.

We should not be surprised by this—if the issues are difficult for medical researchers, they are bound to be difficult for those new to the subject. The statistics are immensely complicated once we get to predicting the consequences for future generations, and the data on which even an expert calculation could be based are incomplete.

The perplexed parent probably can gain enough information to understand the issues if they are persistent. The perplexed parent also needs discrimination—the ability to work out what sources of information are likely to be sound and unbiased. One is likely to come to the wrong conclusion—probably biased in the antivaccination direction—unless discrimination is good. As was argued earlier, the pictures and stories of children who have suffered subsequently to a vaccination are emotionally charged; they are often presented in the absence of pictures and stories of children who have suffered from the disease that the vaccination is intended to eliminate. And they are often promulgated in the absence of statistics about the incidence of vaccination side effects compared with statistics of bad consequences from catching the disease. If the parents include in their calculation the low probability of their own child catching the disease given the generally increased good health of the nation, due to vaccination and/or improved nutrition, they may be sacrificing future generations or their own community. In the case of MMR, these are the kinds of problems parents face, and it does seem irresponsible of the newspapers and a small group of doctors to broadcast the idea that MMR is dangerous in the absence of the other information needed to make a reasonable assessment. Here, then, is a case where science ought to triumph over common sense.

Conclusion

It would be wrong to conclude from this analysis that parents should simply follow the orders of doctors when it comes to vaccination. Medical regimes can be authoritarian and conservative, and they can fail to take on the responsibility of explaining the case to anxious parents in a way that respects their intelligence. But it is equally wrong simply to react against a display of authoritarian behavior. To explore the issues sensibly in the face of the kind of brusque treatment that corrodes trust takes time and expertise; accepting populist accounts of the dangers of vaccines can be to fall into the hands of those with little medical knowledge and much expertise in the art of persuasion. The Web is not controlled; anyone can publish anything.

Science can be wrong—that is the argument of all the volumes of *the Golem*—but this does not make the opposite view right. In the absence of careful research about the opposite view, science is probably the way to bet. This is even more likely to be the case if science is continually put under scrutiny. Here the existence of citizens' groups exploring and pressing cases other than the orthodox view is a good thing so long as it does not turn into anti-science prejudice. Crucially, the citizen must not take the citizens' groups to be right just because they know of cases where authority has been wrong. There will always be occasions when the view from scientific authority will be wrong; that's just the nature of science.

As for the responsibility of scientists, the MMR mess seems to have to do with a medical researcher making a tenuous to non-existent finding public before the real research had been done. Sometimes the authorities overreact to such cases to such an extent that they reduce public trust rather than engender it, but in this case they probably got it right. Andrew Wakefield admitted that he had not proved the link between MMR and autism, yet studies of the making of scientific knowledge, such as we describe in the *Golem* series, show that even a dubious hypothesis can be maintained almost indefinitely and against almost any evidence if its proponent is determined enough. One thing is clear: the public need to understand this; they need to know how to weight anti-establishment scientific opinions and discriminate between kinds of scientist. To understand this, they need to know, not more science, but more *about* science. This will provide not certainty but an input to judgment. The received view may be wrong, and has often been wrong, and it may sometimes favor powerful vested interests, but even this does not make the world as simple as we would like it to be.

Postscript

At the time of writing the latest big epidemiological study, emerging from Denmark, once more exonerates MMR.[16] Perhaps more interesting is a recent attack on vaccination policy from a completely different direction. The claim is that vaccines are preserved with mercury-based compounds, and that the sheer quantity of mercury injected when many vaccinations are given could cause brain damage to young children (mercury preservative is not used in MMR).

There is, perhaps, enough plausibility in that claim to make one think again about whether the precautionary principle is applicable to vaccine programs as a whole. The precautionary principle would suggest less vaccination. We have already argued that "no vaccination" cannot be the appropriate response because of the terrible consequences of many of the diseases which vaccination prevents. But in this case the precautionary principle could be invoked to reduce the numbers of vaccinations. Why vaccinate against chicken pox when there are no well-established serious consequences from catching the disease? Why vaccinate youngsters against influenza, which also does not kill or seriously damage the healthy? When vaccinations are used to reduce inconvenience or financial loss, it may be the moment to say "stop" until the mercury case controversy has had time to reach a consensus among the scientific community.

Late News

On October 7, 2003, the *New York Times* carried an article headed "Refusal of Vaccination Cited in Whooping Cough Outbreak."[17] It describes an outbreak of whooping cough in Westchester County, also in the state of New York, which is blamed

203

squarely on deliberate refusal of parents to vaccinate their children. Measles outbreaks are also being blamed on low take-up of vaccination in various areas of Britain, causing particular grief to parents of children unable to cope with a vaccination because they have another identified ailment.

Conclusion *The Themes Revisited*

We set out on a worried note: *Dr. Golem* was going to be a harder book to write than the other *Golem* books in this series—the books on science and technology. This was first because medicine is much more personal and immediately consequential than the sciences and technologies we examined earlier; however much you try to escape into theory and philosophy, health will get you in the end. How should we react to our friend who, on finding he has tinnitus learns that ordinary medicine can do little for his problem and, trained as a physicist though he was, turns to Chinese medicine and acupuncture for a successful remedy, eventually becoming an advocate for homeopathy? How should we advise our loved one who, on facing the death sentence of a cancer, seeks out alternative care? What do we say to our secretary, who is convinced she has chronic fatigue syndrome, or the colleague who asks for a leave because of his repetitive strain injury? How do we cope with the maelstrom of argument surrounding the vaccinations we are told to administer to our children? How seriously should we take warning signs produced by mammograms, cholesterol counts, and the like? How do we counsel an elderly parent or relative facing the difficult choice of balancing medical intervention with a dignified death? How should we approach our own chronic illnesses and injuries, the ever more eloquent heralds of our own mortality?

Throughout our lives we learn ways to deal with such issues and to navigate their dilemmas. We muddle through, finding the treatment that seems right at the time for us in particular, usually putting aside any deeper reflection on the matter. Because, nevertheless, visits to the doctor can be so significant and because people have been so deeply scarred (sometimes literally) by medical encounters, it is hard to step back as we have tried to do and attempt to offer a dispassionate analysis without running the risk of offending someone.

To make matters worse, health has become a political issue. There are the continued and worsening crises over world health with AIDS, tuberculosis, SARS, or just plain poverty. There is the scandal of the uninsured in the United States. There is the crushing momentum of high-tech, high-cost medicine and the industries that benefit from it—the health insurance companies, the health maintenance organizations, the big drug companies—in the face of the overwhelming evidence that prevention works better than cure. There are the pro-meat and pro-dairy agricultural industries. There is the whole overmedicalization of modern life with its dense population of newly defined diseases, such as attention deficit syndrome. All these have become part of the political as well as the medical landscape. More often than not, talk about the big health issues is talk about politics.

Talk about health in a modern developed society is also talk about consumption and lifestyle. Health is today tied up with identity and the fashion and popular culture industries that thrive upon making the world through advertising. A lifestyle can also be a "health style," whether it be the work-out-every-morning, pill-popping, triple-by-pass business executive, stressing himself in stressing the value of life in the fast lane, or the herbally remedied, health-food loving, organic, yoga-practicing enthusiast for the natural and the calm, who is devoted to avoiding technocratic medical regimes at all costs. Health is never simply health. To deal with health is to assemble and reassemble some of the most important components out of which a modern identity is built.

Our task in this book is also difficult because we have taken an unusual route into matters of health. Because health is so important there are whole academic industries which have taken the analysis of health as their principle activity; there are schools of public health, and the disciplines of health economics, health informatics, and medical ethics. The field we come from, science and technology studies, is, in comparison, tiny. The majority of the studies in the most closely aligned major field—the sociology of medicine—either address only in passing or do not address at all the core issues of this book: What is medical knowledge? What is its relationship to scientific knowledge? How certain is it? Who can possess such knowledge and under what circumstances? and, How much trust should we place in such knowledge? But the marginal location from which we embark perhaps also provides a benefit. As far as we know, no one has treated the problem of medicine exactly as we have here—as a problem of expertise to be looked at in the same way that scholars have looked at the expertise belonging to science and technology. And, as far as we know, this is the first time the crucial distinction for understanding everyday medical decision-making—the difference between succor for the individual and science for collectivity—has been treated as a problem for the sociology of knowledge.[1]

That said, we remind the reader that we too have made choices about what we should assault head-on and what we should skirt around. Readers hoping for solutions to or insights into the large-scale political issues around health will be disappointed. But there are already many books on these matters. Many of the big issues

207

are about resources (lack of them) and resource redistribution. In a way, many of the solutions are obvious; it is just a lack of political will that prevents them from being implemented. We are all for resource redistribution, but this is not a book about it. Only where familiar health policy issues unavoidably touch on what we discuss here have we discussed them or offered our own recommendations. (A clear example is the tension over resources between orthodox and alternative medicine which emerges from the debate about the nature of medical knowledge.)

In *Dr. Golem* we have started with the basic encounter at the core of all health care systems—that between doctor and patient. We have recast this encounter in terms of our theme of expertise. We have tried to chart the contours of expertise as possessed by both patient and doctor. Expertise is a complex matter. It gets more complex when we take into account the distinction between individuals and populations that lies at the heart of the book.

The Contours of Expertise

Given that a visit to the doctor or the hospital can be one of our most consequential interactions with an expert, how should we approach it? We can start by comparing it with other dealings with experts. In the introduction we described a "consultation" with the hairdresser. In that case, if we consider the interaction in sequence, the "patient" has (or should have) the uncontested right over what counts as the illness—bad hair; the hairdresser has uncontested control over how the cure, the hair cut, is managed; and the patient has the uncontested right to say whether a cure has actually been effected. In other words, in this form of expert consultation the customer entirely self-diagnoses the problem; the solution is delivered entirely by the expert; and the evaluation of the solution is entirely in the hands of the customer. The hairdressing consultation is typical of those where the consumer is unambiguously in the driving seat, such as consultations with beauticians, counselors of all sorts, including psychotherapists, and service providers like landscape gardeners.

Other encounters with experts form an inverse (or near inverse) pattern in terms of the contours of expertise. Imagine you want to cure your "disability" in a particular foreign language and consult a language expert. (This is more commonly called "taking language lessons.") The teacher—assuming that he or she is a skilled native speaker in the language to be taught—has uncontested rights over what is to be achieved in terms of your language acquisition, much less in the way of rights over how the teaching is to be carried out (e.g., the student will play the larger role in defining the frequency and duration of lessons, the amount of homework, whether written language skills or conversations skills are required, and perhaps even the style of teaching), but at the end of the "treatment" the teacher is best placed to assess the extent to which a "cure" has been accomplished. This sort of consultation is again typical of a class where the expert possesses such specialized skills that the customer can control only the mode of delivery.

Sometimes this form of consultation is carried out in such a way that the expert also controls the delivery of the expertise—language lessons taken as part of a degree in languages or a modern car brought to a garage for repair. Indeed, with the modern car and the monopolistic garage chain the interaction can be among the most powerless in modern life. The manufacturer's licensed garage takes it upon itself to diagnose the problem, prescribe the treatment, carry out the work behind closed doors, and give the car owner almost no rights to challenge the mechanic as to the effectiveness of the outcome.[2]

Now let us return to the discussion of interactions with medical doctors. How do consultations with medical experts compare with those of hairdressers, language teachers, and car mechanics? As we noted in the introduction, before the growth of medical science in the nineteenth century, interactions with medical experts had

209

more in common with haircutting than with language lessons. Patients played the predominant role in defining their problems and assessing whether a cure had been effected. But over time, as medical science has acquired more diagnostic tools, the interaction has become more like that of the language lesson and, with treatment being delivered in a modern hospital, even like car repair in some respects. One example is the broken limb discussed in chapter 1 and other such gross and obvious injuries. Here medical intervention is at its least perplexing and most one-sided, because the causal chain from intervention to cure is usually easy for the expert (and sometimes patient) to see, and there is usually little uncertainty about outcome.[3] We need a term for this kind of medical intervention to distinguish it from treatments based on averages. In the latter case, the direct causal chains operating in any one individual are unknown or invisible, and the treatment is chosen because it is known to work on average when administered to whole populations as revealed by randomized control trials and the like. So let us refer to treatments of "specific individual causes" (SIC) for things like broken bone repair, and to treatments based on "population average testing" (PAT) for the type of medicine proven by randomized control trials.[4]

Where are we going? Under what we might call "the *Star Trek* model of modern medicine" we will one day have detailed causal explanations for every ailment in every person; the technology of diagnosis and component replacement, from whole organs to single cells, to baths of thought- and mood-affecting chemicals, will be perfected. Under these circumstances, repairing the body will become fully akin to repairing the car, and all PAT will have been replaced by SIC. Right now we live before the *Star Trek* era and we have more PAT than SIC.[5]

In the case of specific individual causes, the consumer's choice when visiting the doctor is relatively straightforward: is the pain and indignity (and perhaps cost), of the intervention adequately counterbalanced by the increase in life quality or expectancy? It is sometimes little more difficult than deciding whether to take your car in for repair when the brake lights fail. You might drive around a bit longer, but your motoring experience and perhaps the longevity of your car (not to mention the risk of being pulled over by the police) dictates that you take it to a garage for repair.

In the case of population average testing, the calculation is more complex because there is always the question of whether the treatment will have any effect on you in particular. For instance, it is known from population testing that, on average, smoking and eating too much butter are bad for you. The consumer can quite reasonably decide that she is the kind of person like her great aunt who smoked a pack a day, drank six gins and tonics every evening, and still lived to be a hundred. Likewise the individual just may be one of those people who are not going to suffer a heart attack as a result of raised levels of cholesterol in his blood even though it is securely known that high cholesterol is correlated with heart disease in whole populations.

One sees, then, that for these sorts of cases of population average testing, the consumer has a reason to decline the expert's advice—or to put it another way, patients have considerable autonomy over their choice of medical intervention in spite of the population results. We may think it foolhardy for someone to limp around for days with what looks like a broken limb without visiting the doctor, but we would be a little less secure in demanding that our aunt give up smoking. In short, individual choice may quite reasonably differ from expert opinion when the expert opinion is too "PAT," if we can exploit the pun.

But now we come to more complex cases, where, even though it is population average testing that is providing the data, the individual choice is not so independent as it seems. This takes us back to the main theme of the book—the tension between individual and collective.

First, consider smoking. If, as modern studies suggest, smoking affects not just you but the health of those around you, the choice is not entirely a matter for the individual, even though the relationship between smoke ingestion and illness is understood only at the level of populations.[6]

The still more interesting cases are where the health of populations has a direct effect on the health of individual. To give a crude but relevant example, if you live on a stream and are stricken with dysentery, the easiest way to remove the infected material from your neighborhood is to void it into the flowing water, protecting your own village and endangering, so it would seem, only those living downstream. This differs from smoking in the following crucial way: in the case of smoking there is no way for the consequences of smoke inhalation by others to effect the original smoker, but if you inflict dysentery on the members of a downstream community, there is a chance that some of them will come and live upstream of you and expose you and your offspring to the danger that you thought you had shifted elsewhere. If we replace the stream of water with the stream of time, vaccination is an exact analogy. If a parent's search for the minimum risk for her child has negative consequences for other children—such as the reintroduction of disease epidemics in the population—as time goes by, these problems "for others" are likely to affect the original child or the child's siblings or offspring. Thus, setting aside all moral considerations, calculations made entirely in terms of short-term self-interest are likely to be incorrect when looked at in the longer term. The analysis applies to all ailments spread by contact with the afflicted.

In these cases, although they are based on PAT, the rights of the consumer to choose the medical intervention and the way it is administered decrease. The calculation of risk is not something that can be done by the individual even if it is only risk to the individual that is at stake. The calculation of risk to the individual is intimately tied to population statistics, and working it out is the prerogative of disinterested specialists; only population statistics reveal the best treatments. Similarly, the ability of the individual to assess the good effects or the damage caused by the intervention is also reduced because, in the current state of medical science, only epidemiology can reveal that there is a causal relationship between a treatment and an ailment (for example, MMR vaccination and autism). Individuals, seemingly gaining relative power from the modern dilution of scientific authority and thinking this gives them much more consumer choice, will make poor judgments in such cases. And this applies even without taking into account the ethics of choice based on individual advantage alone.

VACCINATION PAT AND SIC

We do not live in the *Star Trek* era, and certainly not where vaccination is concerned. The antivaccination pressure group DAN! (Defeat Autism Now!) takes as its motto that "every child is biochemically unique." Under the model of the universe to which a responsible science has to aspire, and which is endorsed in this book, they are right. The consequences are marked. If we consider a case such as MMR, it shows that the epidemiological studies are insufficient to rule out the possibility that a small subset of children with specific biochemical makeup are at risk of becoming autistic in consequence of the jab. There is absolutely no evidence that this can happen, but no epidemiological study can rule it out because epidemiological studies bear on population averages, not specific individuals. What the epidemiological studies do show is that the number of children who are at risk, if any, is too small to show up in the statistics.

But the very logic which shows that there will always be room for worry at the heart of statistical generalizations of this kind also shows why actually worrying is pointless or worse. Why is this? It is because there are an indefinite number of potential worries of

213

this sort. So long as we do not understand the universe exhaustively at the level of specific individual causes, we do not know which of these potentially indefinite number of causes we should worry about, and since we cannot worry about them all, we might as well worry about none of them. For example, suppose I claim that eating kiwi fruit causes the onset of autism. It can be guaranteed that there are a number of children whose autism developed shortly after eating kiwi fruit. To back up the theory we might note that there has been an increase in kiwi fruit consumption in the UK at the same time that there has been an increase in autism. Epidemiological studies might show no correlation between autism rates and kiwi-fruit eating in countries where lots of kiwis are eaten, but these studies cannot prove that there is no small subset of children for whom the effect is real. For kiwi-fruit eating one can substitute almost anything.

That explains why worrying about kiwi-fruit eating, or the indefinite number of equivalents to kiwi-fruit eating, is pointless unless there is some reason for suspicion in addition to either the possibility that it may be true or some observed association in time. Thus, if there were good biological evidence to suggest that kiwi fruits could affect the brain, it would be sensible to stop eating them in spite of a lack of epidemiological evidence. As all the *Golem* volumes show, science is riddled with uncertainties of this sort, but this cannot imply that we must take precautions in respect of everything that has not been proved to be safe or we would, for example, starve to death. The conclusion has to be that we must find a pragmatic path through the issue; the pragmatic path has to be illuminated by what science we have (e.g., epidemiology), because until the *Star Trek* era that's all we have.

Here the guidance we get from the science we have is pretty clear. The epidemiological studies show that failing to vaccinate with MMR is clearly and measurably dangerous just as refusing to eat all the foods that have not been proved to be safe is dangerous.

The point is this: Dr. Wakefield's grounds for saying there is a link between MMR per se and autism is no better than the grounds we have for thinking there is a link between kiwi-fruit eating and autism: some mothers had seen a temporal sequence. (Bear in mind that Wakefield may have had scientific grounds for saying there is a link between measles virus and autism but he continues to recommend measles vaccination, so the only point at issue is the combined vaccine per se.) To draw attention to MMR as an agent which epidemiological studies cannot prove is safe is as arbitrary, then, as choosing kiwi fruit, but in the case of MMR it leads to the unnecessary death or maiming of a number of children.[7]

Finally let us point out that we have given a hostage to fortune. We, nonexperts in the matter of vaccination that we are, have provided a recommendation: "vaccinate with MMR!" On what basis have we done this? It is on the basis of our expertise in the way science works and on what counts as evidence among scientists. Of course, history may prove us wrong. It may turn out that the epidemiological surveys were flawed; it may turn out as PAT turns to SIC that MMR is dangerous. But the fear of giving such a hostage to fortune must not force us to refuse to use our expertise in whatever way we can when we see something going wrong (in this case in the relationship between science and the public). The point is that the parents do not have the luxury of refusing to make themselves into hostages of fortune—they have to decide before history has unfolded, because no action is action in this case: it is the action of exposing a new generation to measles. What we have done is use our expertise to provide clear advice on the basis of the evidence that exists in the middle of 2004.[8]

Note that these arguments apply even though medicine is Golem-like—that is, rough and clumsy. So long as we accept that Dr. Golem has expertise, even if it is analogous to dull industrial diamonds rather than glittering jewels, the types of consideration

215

CONCLUSION

216

of individual choices and the collective good will be the same. The roughness, clumsiness, and dullness make the calculations more inexact but they do not change the direction in which they point.

Part of the roughness and clumsiness arises because not even population effects are known with certainty—randomized control trials can go wrong in many ways, epidemiology is confounded by too many variables, sample sizes are too small, and so forth. The uncertainties at the level of populations are revealed in extreme form in cases such as male circumcision, or tonsillectomy, where medical intervention, purportedly on grounds of health, seems to follow the dictates of medical fashion. For example, with circumcision and tonsillectomy the consumer has very little choice over whether or not there is a problem.[9] The medical profession are doing nearly all the deciding about what is good and bad, even though they change their minds from time to time. Nor is there any choice about process or much to be said about whether the treatment has been efficacious. There is a very strong case for increasing consumer choice in these kinds of interventions. But how does one know ahead of time which are "these kinds of interventions"—the medical-fashion-bound interventions, as opposed to those where there is uncertainty but where the medical profession can legitimately define the problem, diagnosis, and treatment? This problem takes us to the secondary theme of the book—the ways in which we as consumers can increase our knowledge to improve our interactions with medical professionals.

Gaining Expertise

As a first approximation, the more medical understanding a patient has, the better. More understanding of his or her own body helps a patient look for and describe symptoms, improves the taking of the history, and may also encourage more efficient use of medical services. Thus do medical authorities attempt to persuade patients to stop asking for antibiotics to treat viral infections, to stop bothering doctors with trivial complaints, to self-diagnose their own symptoms in case of serious illnesses such as breast cancer or testicular cancer, to understand their bodies sufficiently well to be careful about what they do with orifices and appendages, and to cease ingesting too much in the way of harmful substances, from cigarette smoke to hamburgers and french fries. All this must be a good thing, so long as the illness brought on by reading too much medical literature—hypochondria—does not become endemic.

Medical self-education can also help us choose between experts or challenge a diagnosis. This returns us to medical fashion. If the consumer can discover that medical opinion about some intervention has moved in contrary directions over time, or varies from geographical location to geographical location, then the right to challenge received opinion and the scope for choice must increase. Reading the specialized literature and consulting the Internet are another step on the way to improved medical interaction. Intense discussion with professionals, such as took place in the case of the San Francisco AIDS cure activists, can even lead to a level of understanding which can be called "interactional expertise" in which lay persons, though they may not be in a position to make medical interventions themselves, may be able to make judgments on a par with medical professionals. Finally, in principle, there seems no reason why enough observation and research cannot put the unqualified person in a position to, as we put it, "become a scientist" and contribute to the very definition of new classes of illness.

But, as we have also stressed throughout, the gathering of information should not be mistaken for the acquisition of expertise. The ironic implication of chapter 2, on bogus doctors, is that experience is to be weighted more heavily than information gathering. A bogus doctor with a lot of experience might be better for some purposes than an information-rich young medic fresh out of medical school. Worse, information can become misinformation, especially when the information comes from an unknown source, such as the Internet, where anyone can write anything and give it a gloss of authority. The newspapers and other mass media can also give a completely wrong impression of the force of scientific argument. The media's search for a balanced story can produce a completely wrong impression of the balance of the evidence when the balance of the evidence falls strongly on one side; the case of MMR vaccine in Britain illustrates the point perfectly. Furthermore, an argument is more persuasive when it trades on a sense of immediate danger or, as confidence tricksters know, immediate gain. Often health care decisions in the UK are affected by what health policy experts call "shroud waving"—the media frenzy created by some individual tale of medical woe which subverts the carefully worked out rationality of policy priorities. As we have tried to explain throughout, the search for short-term advantage may not be the best long-term policy even for the purely self-centered individual.[10]

Involvement in discussion groups that include experienced doctors or research scientists can help avoid some of the pitfalls associated with mistaking information for understanding, but the case of the San Francisco AIDS activists shows just how much work and engagement with the medical community is needed to develop the kind of expertise required for lay people to make an intervention in medicine.[11] Chapter 5, which actually examines attempts by lay persons to develop new classes of disease, shows that "becoming a scientist" is very hard indeed. The problem refers to our major theme: though a collection of individuals may be sure that they are experiencing a new syndrome, only technical epidemiological studies along with a disinterested historical perspective can indicate what is likely to be a genuine new disease and what is the overreading of a series of individual impressions which amount to a consumer-led fashion in illness; if medics themselves are subject to fashion, it is hardly surprising that consumers are too.

To summarize, all the medical information that can be gathered is useful in enriching consumer interaction with medical authorities, but the information must be used with humility, never confusing knowledge with expertise. Knowledge is just one component of expertise. Medical training requires doctors, veterinary surgeons, and the like to learn in practice from skilled practitioners—practitioners with experience. This is because many of the skills are tacit and craft-like in character, while book-learning is never enough. Also doctors have to learn how to deal with the uncertainties of medicine in practice, mastering heuristics, rules of thumb, and the other unformalized components of expertise that help them make the correct diagnoses and interventions. If information were the same as expertise, then doctors could be replaced by computers, and, assuredly, they cannot.[12]

Concluding Remarks

An underlying premise of our argument is that it is a good thing when the basis of treatments makes the transition from population average testing to specific individual causes—from PAT to SIC—and it is medical science which enables such transitions to be accomplished. The history of science in the twentieth century and the social studies of science as exemplified in the earlier *Golem* volumes do not encourage us to think that science in general and medical science in particular will ever coincide with the kind of quasi-logical model held dear in naive philosophy and *Star Trek*. If we know one thing with near certainty, it is that the physical world, not to mention the body, will never be entirely understood at the level of detailed causal interactions—in the case of the body, the interactions between cells, chemical messengers, electrical pathways, and thoughts. Nevertheless, we have to hope for more transitions from PAT to SIC. We have to hope for this or we have to embrace an entirely different kind of society where reason is no longer a dominant value, and that is a society that, if we thought about it hard, we would realize we would not like: it would be a society in which, though population statistics for fatalities would no longer work as an argument for universal vaccination, they would also lose their force for seat-belt enforcement, for limiting the use of sport-utility vehicles, for limiting the distribution of guns, for cutting greenhouse emissions, and for reducing smoking in public places. We should also not forget the extent to which the idea of the "natural" has played a central historical role in the politics of racism. But even without a self-conscious rejection of rational and scientific values, it is important to be aware of what can be brought about in spite of the best of intentions. Too vigorous a pursuit of non-science-based therapies at the level of the population may have a secondary effect on medical science. This is a consequence of the economics of scarcity, a force which cannot be forgotten once the gaze is lifted from the individual.[13] We have to decide whether we want the rate of transition from PAT to SIC to slow, stay constant, or speed up. If we do not want it to slow, we must be careful how we play the game of alternative therapy at the policy level.

As for such transitions in the near future, it is not too hard to imagine them coming about for some cancers, and for some brain and nerve damage. In the case of these problems, the time when interventions at the level of an individual's cells can be made, on analogy with intervention at the level of their bones, might be approaching. That said, to repeat the point we argued most forcefully in chapter 4 on vitamin C, it does make perfect sense for the individual to pursue a non-science-based alternative even though nothing follows in terms of collective policy.[14] The logics of medical science and individual treatment are different and the crucial thing is not to mix them up. Medical science should not say "no" to someone who seeks an alternative elsewhere. Too little is known about the body, and certainly too little is known about the interaction of the mind and the body, to justify such a conclusion.

We know there is no ready resolution to the succor versus science, short-term versus long-term, individual versus collective problem, but we have tried to show that short-term solutions cannot be the whole story. More choice for individuals, however compelling these choices may seem to those with only little hope, may mean less choice for others; more choice for this generation may mean less choice for future generations. The answer, insofar as we can supply one, is to make choices with these considerations very much in mind. These choices have to be made in a variety of ways in the context of a variety of levels of knowledge and understanding. The more knowledge and understanding the better, but this does not mean simply accepting what is found on the Internet or in the newspapers; understanding is much harder to attain than that. The medical profession and medical science are bound to make mistake after mistake—that is the nature of science in general and medical science in particular. Medical science will get it wrong even more often than physics or engineering. But it would be incorrect to draw the conclusion that because medical science gets it wrong we should abandon it.

In 1993, in the first volume of the *Golem* series, we wrote:

For some, science is a crusading knight beset by simple-minded mystics while more sinister figures wait to found a new fascism on the victory of ignorance. For others it is science which is the enemy; our gentle planet, our slowly and painfully nurtured sense of right and wrong, our feel for the poetic and the beautiful, are assailed by a technological bureaucracy—the antithesis of culture—controlled by capitalists with no concern but profit. For some, science gives us agricultural self-sufficiency, cures for the crippled, a global network of friends and acquaintances; for others it gives us weapons of war, a school teacher's fiery death as the space shuttle falls from grace, and the silent, deceiving, bone-poisoning, Chernobyl.

In his definitive history of Western medicine, *The Greatest* **221**

Benefit to Mankind, published in 1998, the historian of science Roy Porter expressed a similar sentiment in respect of medicine. He wrote, "For its fans, modern medicine. with its microbe-hunters and micro-chips, has enabled westerners to escape from the valley of the shadow of death, living longer and healthier lives; for critics this is the era of the Holocaust and the Gulags, in whose unspeakable outrages doctors and psychiatrists were hardly reluctant participants. Scientific medicine may be a new knight in shining armor or a new body snatcher" (669).

Our book would have been easier to write if we could have chosen one side or the other. If we could have sided with the doctors, or sided with those who believe that the medical establishment has become "a major threat to health," as Ivan Illich famously claimed in the 1970s, then we could have put forward a livelier and more engaging thesis. We could have concentrated on the elimination, or near elimination, of polio and smallpox, on the way antibiotics have turned killer diseases into inconveniences, and the way death in childbirth has been displaced from its familiar role in drama to an event so rare that it is no longer a credible plotline.[15] Or, we could have pointed out that pharmaceutical products are a leading cause of death in the United States, and that surgeons bungle operations—year-by-year surgeons killed more people in U.S. hospitals during the Vietnam war period than the war itself.[16] But, given what we know today about both sides of knowledge, a book about medicine has to be a less attractive and harder book to write than either of these polemical exercises.

To our surprise, compared to the earlier *Golem* volumes, we find ourselves more on the side of science in this book.[17] In the earlier books we talked of the danger of "flip-flop logic." We felt that science and technology had been oversold; we felt its portrayal of itself as a near divine route to certain knowledge—the flip—would lead to the danger of rejection of science as a whole as its failures to live up to the ideal became evident—the flop. There is too much flop in today's world; it has become all too easy to reject the knowledge of experts and demand it be replaced with a facile populism under the banner of consumer choice. "The knight of medicine," it is true, is no longer dressed in shining armor. The plates that remain creak, rust, and flake; there are jagged edges that can gash and tear; and the sword is dull and chipped. So approach the knight with well-informed caution, but nourish it, polish it, and give it a smile. The knight's mission remains unchanged—to succor those in distress—and the sword still swings.

Notes

PREFACE AND ACKNOWLEDGMENTS

1. Renee Fox provides a systematic discussion of medicine's uncertainties.

2 Just before this book went to press the life of one author's child was saved by emergency surgery following a fall in mountainous country; a ruptured spleen was removed and many units of blood were transfused to replace that lost to internal bleeding, which was on the point of causing irreparable damage and would soon have been fatal. (This footnote is the only alteration made to the book consequent on the accident).

3. Using the approach from Collins and Evans's "Third Wave of Science Studies" paper, we find in medicine a series of inescapable problems to which only an analysis of expertise can provide an answer.

4. Items appearing in the bibliography were significant in writing the text but are not always specifically referenced in the chapters.

INTRODUCTION

1. One minor confusion that we do not deal with occurs when medical scientists announce a "breakthrough," leading sufferers to hope for a new cure when the time between scientific discovery and cure is likely to be decades.

2. The description is based on Epstein's *Impure Science: AIDS, Activism, and the Politics of Knowledge* and is reproduced here by permission of Cambridge University Press. We have chosen to reproduce the chapter exactly as found in *Golem at Large* (with a new introduction) at the risk of a minor disparity in style with other chapters in this volume.

3. Eradication of smallpox was not achieved without cost, it having some of the severest side effects of any vaccination. It has been estimated (in the con-

text of the Iraq war) that a campaign to render the population of the United States freshly immune to smallpox attack would cause thousands of deaths (mostly among the sick, weak, and elderly).

4. The case was complex in that an alternative desired by some parents was a sequence of single vaccinations, although close analysis shows that this option does not alter the principle. The biological connection was said to be between measles virus in the gut and autism, but the measles vaccine would still be administered under the single-shot regime, and parents were advised to continue with single measles shots. Also, though most discussion has been concentrated on measles, the long-term risks carried by rubella and more arguably mumps are thought to be more dangerous than any potential harm emanating from the vaccine. The delay associated with single vaccines, and the added opportunity that this gave for the diseases to spread, therefore imposed more risk, while at the time of writing, the "cocktail" of vaccines per se was associated with no scientific evidence of risk to support the press conference warning, which was not drawn from the paper published by his team in the *Lancet*.

5. It is related to the "The Tragedy of the Commons," according to which if every farmer allows his flock to graze the common land without restriction, then the grass dies and no farmer benefits in the long term. The Prisoner's Dilemma fits a little better in this case, because in the Tragedy of the Commons everyone can see what is going on, whereas in both vaccination and the "Prison" no one knows what choices others are making.

6. Very small numbers of affected cases can get lost in statistical analysis.

7. This is not to say it is easy to persuade parents of the logic of this choice; there can be little worse than agreeing to such an injection and finding oneself soon after with an autistic child. Whatever the logic, the parent is bound to feel guilty.

8. The two authors of this book have both carried out such studies; Collins on human surgery and animal surgery, and Pinch on animal surgery. In the case of both humans and animals the authors witnessed prolonged failures to find veins or organs in the organism undergoing surgery: the cephalic vein was sometimes impossible to find in the case of human pacemaker surgery; a ferret's uterus was not where it should be; and a horse's testicles took half an hour to find. The papers are listed in the bibliography: there is one by Collins, "Dissecting Surgery"; one by Pinch, Collins, and Carbone, "Inside Knowledge"; and one by Collins, Devries, and Bijker, "Ways of Going On."

9. The ideas in this next passage we owe to Jens Lachmund.

10. As we intimate above, we deal here with medicine as it is encountered by relatively well educated people in developed societies. The choices we discuss here are not available in the developing world, nor in areas of developed nations which have a high proportion of very poor or poorly educated inhabitants. In such circumstances there are no choices. We hope that in the fullness of time, everyone will have to face the choices we discuss here.

11. The language of interactional and contributory expertise can be found in Collins and Evans's "Third Wave" paper.

12. For a classic case study of medical training, see Becker's *Boys in White*.

13. Another way of becoming a scientist is described in the AIDS cure study (chapter 7).

14. See Richard Horton, *Second Opinion*. See also his *MMR, Science and Fiction*.

15. Drug companies are the only groups that can afford to finance large numbers of double-blind randomized control trials, and they will run such trials only when there is a chance that the outcome of the tests will yield a return on their investment. See David Horrobin, "Are Large Clinical Trials in Rapidly Lethal Diseases Usually Unethical?"

CHAPTER 1

1. It is obvious in the case of sport that the mental state of the athlete has a very large effect on the level of performance.

2. There could even be a "nocebo effect"—a mentally induced deterioration of health—such as is mentioned in chapter 5.

3. Remember that the existence of expectancy effects and reporting effects has been established by psychologists independently of the existence of the placebo effect proper.

4. This point is owed to Martin Enserink.

5. This point is owed to Andrew Lakoff.

6. This thought is inspired by reports of a double-blind placebo trial which showed no difference between experimental and control groups taking hormone replacement therapy (BBC *Today* program, August 8, 2003).

7. Of the authors of this book, one is sure that he gained immense and immediate relief from a single chiropractic treatment at a time when orthodox medicine could not cure his lower back problem, although he has also tried a number of other alternative treatments for different ailments (including acupuncture), for himself and his children, and found his skepticism markedly increased. None of them worked, nor could he imagine how some of them could possibly have worked. On the other hand, the same author is very critical of a "technology-driven" diagnosis and treatment which failed in the face of what turned out to be readily understood and alleviated symptoms (at the hands of a pharmacist). Those of us who are extremely skeptical when we approach alternative medicine tend to forget the orthodox treatments that fail (as they are bound to); we remember only the alternative treatments that fail. The other author of the book has considerable sympathy for acupuncture.

8. See Smith and Pell, "Parachute Use to Prevent Death and Major Trauma Related to Gravitational Challenge" for a very funny and ironic treatment of the notion that RCTs are appropriate to test all treatments. Our argument goes a little further, separating those cases that should use RCTs and those that

should not into two classes turning on the extent to which the causal chain from treatment to effect is understood at the level of the individual. Clearly the causal chain is well understood in the case of the parachute.

CHAPTER 2

1. There is a strong parallel here with the "experimenters' regress" as discussed in earlier *Golem* volumes. In that case, when the result which a scientific experiment ought to produce is still under discussion, it is unclear whether the experiment has been properly performed.

2. See U.S. House of Representatives, *Fraudulent Medical Degrees, Hearing before the Subcommittee of Health and Long-Term care, of the Select Committee of Aging,* December 7, 1984, 3.

3. *The New York Times,* December 9, 1984, Late City Final Edition, sect. 1, pt. 1, p. 33, col. 1, National Desk

4. In mid-2004, long after the penultimate draft of this chapter had been completed, we came across a much earlier paper on bogus doctors in America—"The Make-Believe Doctors"—which, insofar as it covers the same ground, matches our findings. The paper is by Robert C. Derbyshire and was first published in 1980 and updated in 1990. Derbyshire found evidence for forty-seven impostors between 1969 and 1978 who engaged in serious medical work. Like us, Derbyshire found no women among his American sample, found that there were many bogus doctors and that they could survive for a very long time, and found that they were often unmasked as a result of transgressions having nothing to do with their medical skills. Derbyshire also found that a foothold could be obtained by working in a quasi-medical profession: "By associating with physicians, the impostor learns enough medical jargon to fool the unwary" (46). One way in which Derbyshire has better documentation in his report is in respect of the extent to which patients who have been treated by bogus general practitioners are appreciative of their services even after they are uncovered; we will refer to this again in a later footnote. In his analysis, however, Derbyshire's paper is disappointing. His paper is published by Prometheus Books, and both the book series and the editors of this particular volume are associated, as the book's title indicates, with crude debunking of fringe sciences and the like. Thus Derbyshire is mainly concerned with the scandal of bogus doctors and how they can be reduced in number. He does not look at the general conditions of confidence trickery or the implications for the meaning of modern medicine of the ease with which they can enter and survive in the profession. Nevertheless, his chapter is worth reading for its accounts of bogus doctors' careers. This chapter has been adjusted in the light of our belated discovery of Derbyshire's work only by the addition of this and another footnote.

5. The American cases were collected in 2004 by Matthew Wong. The British research was carried out with support from UK ESRC grant R000234576

entitled "Bogus Doctors: The Simulation of Skill." Joanne Hartland was principal researcher, working with Collins as advisor. The British work was largely completed in 1994 and 1995. Many passages of text found in this chapter are taken directly from early draft work prepared by Hartland with Collins's assistance.

6. For a related case of the assignment of blame, see the events surrounding the Challenger space-shuttle catastrophe; *Golem at Large*, chapter 2.

7. A conservative procedure in terms of what we are trying to find out.

8. There were very few women among the sample—only 5 out of the 87.

9. And the same kind of argument applies to the medical authorities; every reader of this book knows that shampoo is not medicine and so one would not expect a bogus doctor to prescribe it for a throat infection. A bogus doctor, one would think, would be so concerned about keeping his or her reputation intact that shampoo for the throat would be the last thing he would prescribe. Thus, if the shampoo prescription and its stablemates were to be taken as symptoms of anything, it would have been senility, mental instability, or overtiredness, not lack of qualifications. That, perhaps, is why the FPC did not look very hard at Atkins's qualifications in response to the shampoo prescription—it does not look like an action that would emerge out of ignorance.

10. Robert Derbyshire ("Make-Believe Doctors," n. 31) finds that medical impostors, especially in small towns, develop a loyal following who resent the exposure of their trusted family physician. He discussed the case of Freddie Brant, who practiced fraudulently in Groveton, Texas, whose citizens rallied round him when he was unmasked. One farmer said, "My wife has been sick for fourteen years. We've been to doctors in Lufkin, Crockett, and Trinity, and he did her more good than any of 'em. She was all drawed up, bent over, you ought to have seen her. He's brought her up and now she's milking cows and everything" (46). Derbyshire says that juries in the area refused to convict as a result of what the newspaper called a "lava flow" of testimonials supporting him. He also describes as typical the case of the six-year practice of a trusted bogus doctor in a small New York State town: "He even won the esteem of his colleagues who frequently called upon him for consultations. When he was finally exposed . . . the anguished cries of his devoted followers could be heard all the way across the Hudson River. They even circulated petitions to prevent him being banished" (50).

In harmony with the debunking tone of the paper, Derbyshire tries to explain this loyalty as a consequence of what would now be called "cognitive dissonance": the citizens, having accepted their doctors as genuine, did not want to be made out to be fools and so insisted on the value of their services. (Derbyshire concedes, however, that "others may *believe* they have actually been helped"; 51, our stress). We prefer to think that the situation is more as we describe it: given the nature of medical science, there is more than enough room for an empathetic amateur to administer successfully to the majority of regular ailments while referring the difficult cases onward.

11. In other countries, such as Russia and Cuba, medical professionals are paid much less.

12. And with vaccination (see chapter 8) we likewise do not have the information about individuals that can save us from having to rely on population statistics.

CHAPTER 3

1. Information from Joel D. Howell, *Technology in the Hospital.*

2. The significance of these new technologies in increasing the power of the doctor over that of the patient is discussed in the introduction.

3. It is also worth noting that the process of moving from self-diagnosis to consulting a medical expert depends crucially upon factors to do with how a health care system is financed and organized and even the general level of authority accorded medical experts. Issues of access are obviously paramount. If doctors cost money and you have no medical insurance, you are less likely to consult a medical expert. More subtly, if the system has an ethos where wasting the doctor's time is strongly discouraged, you may do more self-diagnosis before consulting an expert. But such a system may also grant too much authority to doctors and put people off from visiting who really need care. Conversely it is well known to doctors in the United States that with patient choice and the lower authority accorded doctors, there are people who come to visit for the most trivial ailments that in other systems may have been dealt with effectively by self-remedy or by a visit to a pharmacist (as is more routine in Germany and Switzerland).

4. Jack L. Paradise et al., "Tonsillectomy and Adenoidectomy for Recurrent Throat Infection in Moderately Affected Children."

5. Of course, there are other conditions for which the operation may be effective, particularly sleep-related breathing disorder (SRBD) and other specific diseases effecting the ear, nose, and throat.

6. Figures come from Jack L. Paradise, "Tonsillectomy and Adenoidectomy," in *Pediatric Otolaryngology,* ed. Bluestone, Stool, and Kenne.

CHAPTER 4

1. See Eric S. Juhnke, *Quacks and Crusaders.*

2. See Collins and Pinch, "The Construction of the Paranormal: Nothing Unscientific Is Happening," in *On the Margins of Science: The Social Construction of Rejected Knowledge,* ed. Roy Wallis, Sociological Review Monographs 27, pp. 237–70 (Keele: University of Keele, 1979), for a detailed examination of the sociology of fringe science.

3. One outcome of Pauling's protest was that the *PNAS* would not in future summarily dismiss such papers and had to give the author a chance to respond to referee comments as with a regular journal.

4. For the same argument worked out in the case of the physical sciences, see Collins, *Gravity's Shadow,* chapter 19.

5. We note, in passing, that Richards's animation—her own energetic defence of Pauling which is exhibited in her book—is informed by her view that scientific testing was not carried out well enough in this case, not that scientific testing was beside the point. Our solution, the one which has informed our arguments throughout this book, is to accept that medical science can never get it right, but can only do its best.

6. Of course, one may not want a medical science at all. One may prefer a return to a more enchanted age which values the "natural" or the magical more highly than the attempt to theorize and measure the efficacy of treatments. But our book is written under the assumption that, for sound or unsound reasons, a medical science is what we should aim for.

CHAPTER 5

1. This reverse placebo effect is known in the literature as the "nocebo effect" (from the Latin "I will harm"). Systematic tests for this effect are rare because of the ethical difficulties of designing a control trial where the desired outcome is to harm patients. There are a few scattered studies supporting the effect, such as one study which showed that women who believed they are prone to heart disease were four times as likely to die as women with similar risk factors. In another experiment asthmatics were told that a harmless vapor they breathed in was an irritant—it produced breathing difficulties in nearly half the patients. These self-same patients were treated for their asthma with a substance they believed was a bronchodilator and they recovered immediately. Although one or two people have suggested in the literature that syndromes such as Gulf War syndrome are manifestations of the nocebo effect, the few researchers interested in the effect mainly want to investigate the possibility of reducing harmful side effects from medications (a proportion of which may be occurring because patients expect the drugs to have harmful side effects).

2. Quoted in Jonathan Banks and Lindsay Prior, "Doing Things with Illness: the Micro Politics of the CFS Clinic."

3. Silverman, "A Disorder of Affect."

4. See Arksey, *RSI and the Experts.*

5. See *Golem at Large.*

CHAPTER 6

1. Stiell et al. "Advanced Cardiac Life Support" (quote on 647).

CHAPTER 7

1. Collins and Pinch, *The Golem at Large,* 126–50.

2. It has been pointed out to us that it should be the subclavian *vein,* not the artery.

1. See the paper by Wakefield et al. in *The Lancet*.

2. See the paper by Wolfe et al.

3. There is an in-between possibility that we initially discussed in chapter 3 on tonsillectomy. A subset of the general population could be more vulnerable to MMR on average. If the members of this subset could be identified, then the risk to them might be found to compare with or exceed the risk from measles even in epidemiological studies. Such a discovery would, of course, be a step on the way to the identification of broken-bone-like causal chains in individuals.

4. We say "almost for sure" because some of the antivaccination Web sites suggest that the eradication of diseases that has normally been credited to vaccination is actually due to improvements in health and nutrition. (It seems hard to credit that this is the cause where diseases have been eliminated or much reduced in prevalence in developing countries.)

5. It has been argued that rubella vaccine should not be administered, but that rather all responsible parents should make certain that their girl children catch rubella at a young age to make sure they cannot fall prey to it during pregnancy (with severe risk to the baby). The argument is that antirubella vaccination is not as good a safeguard against disease as catching the disease itself. The logic of this is that it should be the government's responsibility to actively infect young girls with rubella. The argument does not affect the overall logic of the case we are making here, however, even if it might affect the detail. Before the rubella vaccination it was certainly the habit of parents (at least in Britain) to put their young girl children in contact with someone infected with rubella.

6. James Naughty chaired the discussion.

7. Pinch, who was in the United States during this period, was not privy to the debate in the UK and did not hear the Labour spokesperson's broadcast. Nevertheless he believes that the Labour Party's resistance to the separate vaccine policy is based on the belief that it would give rise to a lower take-up of the less deadly elements of the cocktail of vaccines, namely mumps.

8. So far a hundred or so.

9. We are grateful to Robert Evans for stressing this point.

10. Writing in late 2003, Collins's position is in the minority, at least among academics, whereas Pinch's view is widely accepted. Collins's view is worked out and defended in the papers by Collins and Evans, "The Third Wave" and "King Canute." Note that the case might be different if Wakefield's view had a substantial body of research, even it were a minority view, to back it up.

11. Of particular relevance was DTP 10/15/91 issued by the Department of Health and Human Services, Centers for Disease Control, Atlanta, Georgia, and "Immunisation Review," issued by Connaught Laboratories, Inc., which was not dated.

12. There appears to be some mistake, as Pinch has quoted a leaflet previ-

ously saying that the figure is 1 in every 1,750, but Collins did not spot this at the time [HMC].

13. But see Barry Glassner's *The Culture of Fear;* he documents the way the pertussis scare spread in the United States and argues that the new vaccine is both more expensive and less effective (174–79).

14. The case, though the Pinches did not know this at the time, is slightly complicated because herd immunity is very difficult to bring about in the case of pertussis. This is because the vaccine is not 100 percent effective; its effects wear off after a few years; and older people are not given the vaccine, as the side effects are worse for older people. Hence the disease remains endemic but can be kept to a low level in the group most at risk from its consequences (the young) by vaccinating that sector of the population even though they could still catch it from parents and older siblings in later life.

15. Our thanks to the administrative staff of the Cornell Program in Science and Technology Studies for locating a newspaper article which led us to this data.

16. See the paper by Maadsen et al.

17. The author is Richard Perez-Pena.

CONCLUSION

1. The individual-collective tension is, of course, a staple of social policy—see Richard Titmuss's famous comparative analysis of national systems of blood donorship and transfusion, *The Gift Relationship.*

2. Something which should be a cause of concern for the consumer organizations.

3. Actually, even here things are not entirely unproblematic: evidence from the Falklands War suggests that badly wounded soldiers who were left out in the cold overnight rather than given immediate emergency treatment had a higher recovery rate. The argument is that lying out in the cold slows the body's processes and allows internal wounds to form scabs, sealing blood loss, whereas movement and immediate restoration of blood pressure through the administration of fluids prevent internal vessels from sealing themselves (but see preface, note 2). Of course, things go badly wrong when problems are treated as though they can be resolved by reference to a "fault tree" when individual variation remains important and poorly understood.

4. Like nearly all such dichotomies there is a degree of overlap. Thus, though one may fully understand the causal chains involved in, say, heart transplants, or mastectomy, a true picture of how much they increase life expectancy requires averaging their effects over all operations and comparing with noninterventions.

5. See chapter 8, note 5, for a discussion of an in-between case.

6. We neglect more subtle effects such as the visible example of your smoking encouraging others to smoke and your smoking helping to support

the tobacco industry and thus indirectly providing more opportunities for others to smoke.

7. Interestingly, tree nut allergy causes well over 100 deaths per year in the United States, but nobody suggests a universal ban on tree nuts. This makes the intensity of feelings over MMR, which does, after all, do a lot of good, still more pointed.

8. For a more systematic treatment of the way the speed of politics, or in this case, parental choice, exceeds the speed of science, see Collins and Evans's "Third Wave" paper.

9. With the exception of certain religious groups.

10. Even when we turn to the peer-reviewed journals we cannot simply rely on what we read. In spite of the positive rhetoric associated with peer review, there are many mistakes in technical medical papers in well-monitored publication outlets; only long experience helps distinguish the secure from the dubious when reading even the mainstream journals. In the book *Gravity's Shadow*, Collins shows that even in the physical sciences, journal articles cannot be taken at face value; some published papers, which are, on the face of it, of enormous importance, are routinely ignored by scientists who are "in the know."

11. There is a large literature on the difference between garnering information and developing skills and understanding that is forgotten over and over again in what sometimes seems like a systematic way. Thus, it is in the interests of those who want to make education less costly to forget the difference between expertise and information so that mass distance learning can replace experienced and highly paid teachers. At the other end of the political spectrum, it is in the interests of those who favor the democratization of all knowledge to forget the difference so that they can claim that it is easy for lay persons to attain the understanding needed to challenge professionals by reading and the like, without needing to be trained by experts.

12. Collins's *Artificial Experts* explains some of the reasons why this is so.

13. This does not mean we endorse the wholesale application of economic thinking to health. for criticism of health economics, see Ashmore, Mulkay, and Pinch, *Health and Efficiency*.

14. This is not to say that we do not need some continued *scientific* investigation of alternatives, but that is not the same as allowing consumers to choose the direction medicine should take.

15. The mother dies in only one in 10,000 of today's births in Western countries.

16. On pharmaceutical products and medical interventions as a leading cause of death, see Hasslberger, "Medical System Is Leading Cause of Death and Injury in US." The figures on deaths from surgery come from a 1974 Senate investigation quoted in Roy Porter, *The Greatest Benefit to Mankind*, 687.

17. Those volumes were on the side of science too, though our critics, and some of our colleagues, failed to notice it.

Bibliography

Arksey, Hilary. *RSI and the Experts: The Construction of Medical Knowledge*. London and Bristol, Pa.: UCL Press, 1998.

Aronowitz, Robert A. *Making Sense of Illness: Science, Society, and Disease*. Cambridge and New York: Cambridge University Press, 1998.

Ashmore, Malcolm, Michael Mulkay, and Trevor Pinch. *Health and Efficiency: A Sociology of Health Economics*. Milton Keynes: Open University Press, 1989.

Baker, Jeffrey P. "Immunization and the American Way: 4 Childhood Vaccines." *American Journal of Public Health* 90.2 (2000): 199–207.

Banks, Jonathan, and Lindsay Prior. "Doing Things with Illness: The Micro Politics of the CFS Clinic." *Social Science and Medicine* 52 (2001): 11–23.

Becker, Howard S. *Boys in White: Student Culture in Medical School*. Chicago: University of Chicago Press, 1981.

Beecher, H. K. *Measurement of Subjective Responses*. New York: Oxford University Press, 1959.

Beecher, H. K. "The Powerful Placebo." *Journal of the American Medical Association* 159 (1955): 1602–6.

Blaxter, Mildred. "The Cause of Disease: Women Talking." *Social Science and Medicine* 17 (1983): 59–69.

Bloor, Michael. "Bishop Berkeley and the Adeno-tonsillectomy Enigma." *Sociology* 10 (1976): 43–61.

Bosk, Charles L. *Forgive and Remember: Managing Medical Failure*. Chicago: University of Chicago Press, 1979.

Brown, Phil. "Popular Epidemiology and Toxic Waste Contamination: Lay and Professional Ways of Knowing." *Journal of Health and Social Behavior* 33 (1992): 267–81.

236

Brown, Phil, et al. "A Gulf of Difference: Disputes over Gulf War—Related Illnesses." *Journal of Health and Social Behavior* 42 (2000): 235–57.

Bynum, W. F., C. Lawrence, and V. Nutton. *The Emergence of Modern Cardiology.* London: Wellcome Institute for the History of Medicine, 1985.

Cameron, Ewan. *Hyaluronidase and Cancer.* New York: Pergamon Press, 1966.

Collins, H. M. *Artificial Experts: Social Knowledge and Intelligent Machines.* Cambridge, Mass.: MIT Press, 1990.

Collins, H. M. "Dissecting Surgery: Forms of Life Depersonalized." *Social Studies of Science* 24 (1994): 311–33.

Collins, H. M. *Gravity's Shadow: The Search for Gravitational Waves.* Chicago: University of Chicago Press, 2004.

Collins, H. M., G. Devries, and W. Bijker. "Ways of Going On: An Analysis of Skill Applied to Medical Practice." *Science, Technology, and Human Values* 22.3 (1997): 267–84.

Collins, H. M., and Robert Evans. "King Canute Meets the Beach Boys: Responses to the Third Wave." *Social Studies of Science* 33.3 (2003): 435–52.

Collins, H. M., and Robert Evans. "The Third Wave of Science Studies: Studies of Expertise and Experience." *Social Studies of Science* 32.2 (2002): 235–96.

Collins, Harry, and Trevor Pinch. *The Golem: What Everyone Should Know About Science.* Cambridge and New York: Cambridge University Press, 1993. [2nd ed. in paperback; Canto, 1998]

Collins, Harry, and Trevor Pinch. *The Golem at Large: What You Should Know about Technology.* Cambridge and New York: Cambridge University Press, 1998. [Paperback ed.; Canto, 1998]

"Complementary Medicine." Special issue, *New Scientist* 2292 (May 26, 2001).

Derbyshire, Robert C. "The Make-Believe Doctors." In *The Health Robbers: A Close Look at Quackery in America,* edited by Stephen Barret and William T. Jarvis, 45–54. Buffalo: Prometheus Books, 1980. [2nd updated version 1990]

Enserink, Martin. "Can the Placebo Be the Cure?" *Science* 284 (1999): 238–40.

Epstein, Steven. *Impure Science: AIDS, Activism, and the Politics of Knowledge.* Berkeley, Los Angeles, and London: University of California Press, 1996.

Fox, Renee C. "Medical Uncertainty Revisited." In *Handbook of Social Studies in Health and Medicine,* edited by Gary L. Albrecht, Ray Fitzpatrick, and Susan C. Scrimshaw, 409–25. London, Thousand Oaks, and New Delhi: Sage, 2000.

Friedman, N. *The Social Nature of Psychological Research.* New York: Basic Books, 1967.

Glassner, Barry. *The Culture of Fear.* New York: Basic Books, 1999.

Groopman, Jerome. "Hurting All Over." *New Yorker,* November 13, 2000, 78–92.

Hardy, Michael. "Doctor in the House: The Internet as a Source of Lay Health Knowledge and the Challenge to Expertise." *British Medical Journal* 321 (1999): 1129–32.

Harrington, Anne, ed. *The Placebo Effect: An Interdisciplinary Exploration*. Cambridge Mass.: Harvard University Press, 1997.

Harrow, David H. "Indications for Tonsillectomy and Adenoidectomy." *Laryngoscope* 112 (2002): 6–10.

Hasslberger, Josef. "Medical System Is Leading Cause of Death and Injury in US." *NewMediaExplorer*. http://www.newmediaexplorer.org/sepp/2003/10/29/medical_system_is_leading_cause_of_death_and_injury_in_us.htm.

Helman, Cecil G. "'Feed a Cold, Starve a Fever': Folk Models of Infection in an English Suburban Community, and Their Relation to Medical Treatment." *Culture, Medicine and Psychiatry* 2 (1978): 107–37.

Horrobin, David F. "Are Large Clinical Trials in Rapidly Lethal Diseases Usually Unethical?" *Lancet* 361 (February 22, 2003): 695–98.

Horton, Richard. *MMR: Science and Fiction; Exploring a Vaccine Crisis*. London: Granta Books, 2004.

Horton, Richard. *Second Opinion*. London: Granta Books, 2003.

Howell, Joel D. *Technology in the Hospital: Transforming Patient Care in the Early Twentieth Century*. Baltimore: Johns Hopkins University Press, 1995.

Hrobjartsson, Asbjorn, and Peter C. Gotzsche. "Is the Placebo Powerless? Analysis of Clinical Trials Comparing Placebo with No Treatment." *New England Journal of Medicine* 344 (2001): 21, 1594–1602.

Illich, Ivan. *Limits to Medicine: Medical Nemesis, the Expropriation of Health*. Harmondsworth: Penguin, 1976.

Juhnke, Eric S. *Quacks and Crusaders: The Fabulous Careers of John Brinkley, Norman Baker, and Harry Hoxsey*. Lawrence: University Press of Kansas, 2002.

Lachmund, Jens. "Between Scrutiny and Treatment: Physical Diagnosis and the Restructuring of 19th Century Medical Practice." *Sociology of Health and Illness* 20 (1998): 779–801.

Lachmund, Jens. "Making Sense of Sound: Auscultation and Lung Sound Codification in Nineteenth-Century French and German Medicine." *Science, Technology and Human Values* 24 (1999): 419–50.

Lachmund, Jens, and Gunnar Stollberg, eds. *The Social Construction of Illness: Illness and Medical Knowledge in Past and Present*. Stuttgart: Franz Steiner Verlag, 1992.

Lakoff, Andrew. "Signal and Noise: Managing the Placebo Effect in Antidepressant Trials." Paper presented to meeting of the Society for Social Studies of Science. Cambridge, Mass., November 1–4, 2001.

Lloyd, Andrew R. "Muscle versus Brain: Chronic Fatigue Syndrome." *Medical Journal of Australia* 153 (1990): 530–33.

Maadsen, Kreesten, et al. "A Population Based Study of Measles, Mumps, and Rubella Vaccination and Autism." *New England Journal of Medicine* 347.19 (2002): 1477–82.

Mathews, J., et al. "Guillotine Tonsillectomy: A Glimpse into Its History and

Current Status in the United Kingdom." *Journal of Laryngology and Otology* 116 (2002): 988–91.

Maurer, D. W. *The Big Con: The Story of the Confidence Man and the Confidence Game*. New York: Bobbs Merrill Co., 1940.

Millman, Marcia. *The Unkindest Cut: Life in the Backrooms of Medicine*. New York: Harper, 1976.

Monaghan, Lee F. *Bodybuilding, Drugs, and Risk*. London and New York: Routledge, 2001.

Paradise, Jack L. "Tonsillectomy and Adenoidectomy." In *Pediatric Otolaryngology*, edited by C. D. Bluestone, S. E. Stool, and M. A. Kenne, 1054–65. Elsevier Science Health Science Division, 1996.

Paradise, J. L., et al. "Efficacy of Tonsillectomy for Recurrent Throat Infection in Severely Affected Children: Results of Parallel Randomized and Non-randomized Clinical Trials." *New England Journal of Medicine* 310 (1984): 674–83.

Paradise, J. L., et al. "Tonsillectomy and Adenoidectomy for Recurrent Throat Infection in Moderately Affected Children." *Pediatrics* 110.1 (2002): 7–15.

Pauling, Linus. *Vitamin C and the Common Cold*. San Francisco: W. H. Freeman, 1970.

Pinch, T., H. M. Collins, and L. Carbone. "Inside Knowledge: Second Order Measures of Skill." *Sociological Review* 44.2 (1996): 163–86.

Porter, Roy. *The Greatest Benefit to Mankind*. New York: W. W. Norton and Co., 1998.

Porter, Roy. *Health for Sale: Quackery in England, 1660–185*. Manchester and New York: Manchester University Press, and St. Martin's Press, 1989.

Prior, Lindsay. "Belief, Knowledge, and Expertise: The Emergence of the Lay Expert in Medical Sociology." *Sociology of Health and Illness* 25 (2003): 41–57.

Richards, Evelleen. *Vitamin C and Cancer: Medicine or Politics?* London: Macmillan, 1991.

Rosenberg, Charles E. *The Cholera Years: The United States in 1832, 1849, and 1866*. Chicago: University of Chicago Press, 1962.

Rosenberg, Charles E., and Janet Golden, eds. *Framing Disease: Studies in Cultural History*. New Brunswick, N.J.: Rutgers University Press, 1992.

Rosenthal, Robert. "Interpersonal Expectancy Effects: The First 345 Studies." *Behavioural and Brain Sciences* 3 (1969): 377–415.

Rosenthal, Robert. "Interpersonal Expectations." In *Artifacts in Behavioural Research*, edited by R. Rosenthal and R. C. Rosnow. New York: Academic Press, 1978.

Rosenthal, Marilyn M. *The Incompetent Doctor: Behind Closed Doors*. Buckingham: Open University Press, 1995.

Shapiro, Arthur K., and Elaine Shapiro. "The Placebo: Is It Much Ado about Nothing?" In *The Placebo Effect: An Interdisciplinary Exploration*, edited by

Anne Harrington, 12–36. Cambridge, Mass.: Harvard University Press, 1997.

Silverman, Chloe. "A Disorder of Affect: Love, Tragedy, Biomedicine, and Citizenship in American Autism Research, 1943–2003." PhD diss., University of Pennsylvania, 2004.

Singer, M., et al. "Hypoglycemia: A Controversial Illness in US Society." *Medical Anthropology* 8 (1984): 1–35.

Smith Gordon, C. S., and Jill P. Pell. "Parachute Use to Prevent Death and Major Trauma Related to Gravitational Challenge: Systematic Review of Randomized Control Trials." *British Medical Journal* 327 (2003): 1459–61.

Stiell, Ian G., et al. "Advanced Cardiac Life Support in Out-of-Hospital Cardiac Arrest." *New England Journal of Medicine* 351 (August 12, 2004): 647–66.

Stolberg, Sheryl Gay. "Sham Surgery Returns as a Research Tool." *New York Times,* April 25, 1999.

Talbot, M. "The Placebo Prescription." *New York Times Magazine,* January 9, 2000. Accessed at http://www.nytimes.com/library/magazine/home/20000109mag-talbot7.html.

Thornquist, Eline. "Musculoskeletal Suffering: Diagnosis and a Variant View." *Sociology of Health and Illness* 17 (1995): 166–80.

Timmermans, Stefan. *Sudden Death and the Myth of CPR.* Philadelphia: Temple University Press, 1999.

Titmuss, Richard. *The Gift Relationship: From Human Blood to Social Policy.* Harmondsworth: Penguin, 1973.

Wakefield, A. J., et al. "Ileal-Lymphoid-Nodular Hyperplasia, Non-Specific Colitis, and Pervasive Developmental Disorder in Children." *Lancet* 351 (1998): 637–41.

Watts, Geoff. "The Power of Nothing." *New Scientist* 2292 (2001): 34–37.

Wolfe, Robert M., Lisa K. Sharp, and Martin S. Lipsky. "Content and Design Attributes of Antivaccination Web Sites." *Journal of the American Medical Association* 287.24 (2002): 3245–48.

Wright, P., and A. Treacher, eds. *The Problem of Medical Knowledge: Examining the Social Construction of Medicine.* Edinburgh: University of Edinburgh Press, 1982.

Index